Vincent Scoper, Jr.

P.O. Box 7998
Tyler, TX 75711

Come Drill A Well In My Back Yard

Printed in the United States of America

First Printing May, 1971
Second Printing April, 1972
Third Printing Jan., 1973
Fourth Printing Oct., 1973
Fifth Printing Dec., 1974
Sixth Printing Sept., 1976
Seventh Printing Jan., 1978
Eighth Printing July, 1978

CONTENTS

Chapter 1

What this book is all about

"There's oil on my land!! . . . It's coming up out of an old water well and flowing all over the place!!" These were the excited statements that I heard when I answered the telephone early one morning. I finally calmed the speaker long enough to find out his name and some of the facts about his "oil well." It seemed that he had bought a farm a few weeks before and had just found an abandoned water well that he didn't know existed. The rusted pipe was sticking out of the ground and oil was flowing over the rim.

I didn't think there was a chance in a million that he actually had oil flowing out of a water well, but you never can tell . . . so I told him I'd come take a look. He was waiting for me at his front gate. Before we had taken a dozen steps, he was telling me that I ought to drill at least 10 wells on his land, each of which should produce 100 barrels of oil per day. Even though I knew there was probably nothing to this, he was so excited that some of his enthusiasm started to rub off on me. I began to feel a tingle of excitement; maybe he did have something.

My excitement was short-lived however, because this "flowing oil well" turned out to be rain water that had collected in a puddle around the well. The "oil" was a rainbow film on top of the water. It was probably caused by decaying vegetation in the area. After hearing his glowing report about oil flowing out of a pipe in the ground, my first impression was that he had brought me to the wrong place. The only resemblance to what he had told me and what was actually there was the rusted water pipe sticking out of the ground. Since he had talked himself into believing that oil was on his land, no amount of explaining could change his mind. I left knowing that he felt I had some secret knowledge that I would not share with him.

A variation of this story has happened many times, but the "oil" was **always** found on the land of the individual who called me. It seems that no one has ever found any "oil" on someone **else's** land. Everyone, I suppose, has had the dream of finding oil on his land. And this thought brings us to the title of this book.

For the past fifteen years, I've heard this statement dozens of times: "Come drill a well in my back yard." The person asking the question—and it is a question—is usually half serious—half joking, not really knowing whether or not I can take a look at his land and tell if there is any oil below. I don't mind discussing "oil on his land," however, because in explaining why I can't help him by just looking, we usually end up discussing the oil business from "how oil is formed" to how leases, minerals and royalty are bought and sold. This I enjoy.

As a result of many discussions on this subject, I have come to realize that most people are eager to learn of the many fascinating facts about oil. Not that I personally am so gifted as to make an uninteresting subject interesting—I just state the facts as they are and it turns out interesting all by itself. Nor do I presume that these writings will add much to the world of literature of journalism but when you finish reading this book, you should have a pretty fair working knowledge of the oil business. Perhaps the information contained within will help you make or save a dollar or two or three or more.

In writing this book, I have tried to discuss every type of question that has ever been asked me. The following are a few of the most recurrent questions. These questions, plus many more, are discussed and explained in detail:

How much money can I expect every month if they strike oil on my land?

How was oil formed and how did it get under the ground?

Is an oil pool really a pool under the ground?

Do "gushers" still occur?

Could I really drill a well in my back yard if I wanted to?

Would an oil company drill a well, cap it as a dry hole, and then come back years later to produce it?

What is "making a trip"?

Can a geologist just look at the land and tell if there is oil below?

What is "bringing in a well"?

What is a "christmas tree"?

What is "fishing"?

What is a "wildcat well"?

Do you really know if oil is there before you drill?

Once you hit salt water in a well, can you still find and produce oil?

Why does my neighbor get part of the oil when the well is on my land?

The questions that are asked are really divided into two categories: The first, as the questions above, could be classed as general knowledge or curiosity questions. The second category are questions concerning how money is made in oil. The following are a few of the questions of this type:

How can a person not connected with the oil business invest in oil?

How does an oil lease work?

What are minerals and royalty?

You've got it made if you have an interest in a producing oil well, don't you?

How much risk is involved in investing in the oil business?

If I invested $1000.00 in a well and it produced, how much return on my investment could I expect?

Is there really a tax advantage in the oil business?

In the first part of this book I will try to acquaint you with a general knowledge of oil production; while in the last chapter, I will try to explain the potential and the risk involved in investing in the oil business.

This book should appeal basically to three groups of people: (1) Those who live in an oil producing area or a potential oil producing area who might have their land leased already or might even have producing wells on their land. (2) Those who would like to invest in the oil business but don't know the first thing about it. (3) Those who are just curious about an interesting subject. This could apply to someone who might be presently working in some phase of the oil business but would like to know more about all phases of the business.

This is not intended to be a technical book. It is written so that anyone can understand and, I hope, enjoy it. To get statistics and figures alone about oil, consult an encyclopedia; statistics, etc., will be used sparingly and only when necessary to help you understand a particular point.

I'll explain how oil was formed, how it is found, how a well is drilled, how the oil is produced, and how to invest in oil if you have a little gambling blood in you. If you've never experienced the sensation, few things can equal the stomach-churning anticipation of waiting for a well to be drilled in which you own an interest.

Chapter 2

Two illogics do not make one logic

I was conducting oil exploration for a major oil company in south Florida some years ago and by the nature of my work, it was necessary to talk with the landowners on many occasions. Of the hundreds that I have talked with concerning the prospects of finding oil on their land, all but one "knew" there was oil below. Of course it was 99 percent wishful thinking, but every once in a while an individual will have a legitimate reason for thinking that he might have oil on his land. If his reasoning is valid, he might be able to convince someone to drill a well.

In most cases, however, the concept that people have about how oil is found under the ground never ceases to amaze me. I'd like to be quick to point out that this statement is not said in criticism of the people who told me about their ideas. These concepts involve original thinking and, oddly enough, some of them sound very logical if an individual is not aware of the facts.

Pools of oil?

Many people think that oil is found in huge cave-like cavities beneath the surface. They assume that you just drill a hole and suck it out like a straw in a soda. After all, it's called an oil pool isn't it? This concept is completely wrong, but I'm afraid that people who work in the oil business who know better have given the general public this erroneous idea. It is called an oil pool, even by oil people, but it is not found in a pool. This will be discussed in detail later.

Rivers of oil?

Another popular misconception is that oil is found in rivers under the ground. The oil is thought to be flowing along from one place to another and if you don't drill at the right place at the right time, the oil will be gone. This also is not true, but one elderly gentleman in Florida needed more convincing than my word on the subject. He told me that a river of oil started at the northern end of Florida, flowed down the center of the state, and out into the Gulf of Mexico. Since I had never heard this one before, I was anxious to know how he had arrived at this conclusion. Looking at me as though I had just asked him what two plus two was, he very condescendingly said, "Why, all I have to do is put my ear to the ground on a still night and I can hear it flowing along." Whatever he heard, it wasn't oil, because oil doesn't flow in rivers under the ground. I tried every way I could think of to convince him that this was not the situation, but he had made up his mind and did not want to be confused by the facts. In parting he said, "For a geologist, you sure don't know much about oil, do you?" And I mumbled, "No, I guess not."

There ain't no oil in Florida

Only once did I meet someone who did **not** think there was oil under his land. In fact, he went so far as to say that there was no oil in the entire state of Florida.

Sometimes a person's belief concerning certain aspects about oil is so far removed from the facts that a sensible discussion is almost impossible. With all due respect to wives, you men can understand the point I'm trying to make. When you sometimes have a difference of opinion with your wife and try to "discuss" this difference with her, you get the feeling that the two of you are talking about a completely different subject. Her illogic has so many illogical points that it actually begins to sound logical. But just as two wrongs do not make a right, two illogics do not make a logic.

Well anyway, the company for which I worked had had a producing field in Florida for the past 14 years, very close to this doubting Thomas' home. I felt that I finally had something concrete with which to make a point. I reminded him that there was a producing oil field less than six miles from his house and he could see the storage tanks and other producing equipment from the road that he traveled every day.

"Well, I've seen the storage tanks, but I ain't seen no oil. They just got all that stuff there to make people think there's oil being produced." I never could quite understand what he thought the purpose of this "deception" was, but I explained that he didn't see the oil because it came out of the ground and went directly into the storage tanks. From there, it was pumped into trucks, taken across the state, and then transferred into a tanker for shipment up the east coast.

I thought to myself, "he can't possibly find anything wrong with this explanation." He looked me over rather cautiously and said, "Son, you look like an honest person . . . I want to ask you a question and I want an honest answer." I shuddered a little at what might be coming but I agreed that honesty is the best policy. "Have you ever actually **seen** any of that oil with your own two eyes?"

As I look back, I should have told something less than the full truth and said "yes." But what can you do when you've just been tagged as an honest person! After a long pause I reluctantly said, "Well, no, I actually haven't seen the oil with my own two eyes, but. . ."

"Aha! you work for the company and they've even got you fooled!" For some reason, the discussion abruptly turned to alligators and the Everglades.

Another misconception is the idea that if you own an interest in an oil well, you have it made—not true! Did you know that owning an interest in a producing oil well could actually cost you money? There is never any thought given to how much it costs to produce the oil, how much oil is produced, or how much interest is owned. There is just one "logical" thought . . . oil means wealth. This is another one of those "logical" thoughts that turns out to be illogical.

If a well produces an income of $1,000.00 per month but the cost of producing, such as electricity or gasoline to operate the pumping unit, etc., is $1,500.00 per month, you can readily see that it is actually costing $500.00 per month to produce the well. Of course, if an individual owns royalty in a well, he doesn't have to share in the cost of producing the well. He just gets a check every month. This is discussed more in detail later, but very often a well is plugged and abandoned while it is still producing several barrels of oil per day.

It is a very difficult decision to plug a well which is still producing even a few barrels of oil per day. Even though it's a losing proposition, the operator keeps thinking that if he could only keep the operating expenses down, he could start making a profit again. This can go on and on.

There is the story of a man who owned interests in several poor wells—the monthly operating expenses were about the same as the amount of money that the oil was bringing—and he was asked what he would do if he had a million dollars. He thought for some time, replying as if he had no choice in the matter "Well, I guess I would just stay in the oil business until it was all used up." Of course this is not the rule, but the exception. The only reason for making this point is to say that not everyone makes money in an oil venture. Needless to say, there have been many fortunes made from oil, but many have been lost also.

These are just a few of the misconceptions about oil that I have run into. If you have any preconceived ideas about oil and the oil business, this book, read with an open mind, should prove time well spent.

Chapter 3

How oil is formed

In order to understand where oil came from, how it accumulated in one place, and how to look for it, it is necessary to realize how very, very old this earth is and how many physical changes have taken place. We do not know exactly how old the earth is, but we do know that it is at least 5 billion years old because we've found rocks that are that old. How do we know they're 5 billion years old? There is a scientific method of dating rocks and fossils that measures radioactive decay of the material being studied. Basically it measures the number of radioactive particles given from a substance during a certain length of time. This method of telling the age of rocks and fossils is authentic and surprisingly accurate.

How long is 5 billion years?

It's difficult for the mind to imagine how long 5 billion years is, but here is a presentation that will help you to understand and also stagger your imagination. Mentally condense 5 billion years into one year, from January to December—this is a length of time we can easily comprehend. If the earth were created on January 1, at what time of the year would you guess that man first appeared? October? . . . No! November? . . . No! Man appeared during the last day of the year, the last hour of the day, the last minute of the hour and the last 30 seconds of the last minute of the year. When you think of the countless volumes of history we have recorded within the past 3000 years, and realize that all recorded history is just an instant in the life of our earth, you can begin to understand how long 5 billion years is.

I feel that at this point I should make a brief statement relating theology to the age of the earth and to when man first appeared: I believe that God created the earth and I believe that He created man. The manner, or the length of time, in which he chose to accomplish this, I do not question . . . nor does it lessen my faith in Him. In fact, when I realize how our earth and entire universe can be so complex and yet at the same time, so orderly, it cannot help but strengthen my faith in God.

We measure the geologic past by means of a Geologic Time Chart. Each interval of time has been given a name so that a particular time in the past can be referred to more easily. In recent times, we refer to periods in history with terms such as the Ice Age, Iron Age, and Atomic Age. Since we are dealing with millions of years when speaking of the geologic past, we use different terms. As an example: dinosaurs became extinct approximately 70 million years ago. Another way of saying it . . . dinosaurs died out at the end of the Cretaceous Period. Incidentally, the stereotyped movies of cave men and dinosaurs engaging in mortal combat is historically impossible. Dinosaurs became extinct millions of years before man appeared on earth. These names are also used when referring to oil producing formations. Should you hear someone say that oil is being produced from the

Cretaceous, you know that oil is being produced from rocks that were formed over 70 million years ago. (Figure 1)

During these 5 billion years seas and mountains have come and gone and there have been drastic temperature changes. One evidence of this is the fact that fossils of palm trees have been found at the North Pole. This tells us that the Arctic once had a warm climate in the geologic past. Shark teeth have been found in areas that are hundreds of miles from a present body of salt water and thousands of feet above present sea level. In fact, many sections of the United States have been covered by the sea and risen up again several times.

Any time a part of the United States was under water, then of course rivers flowed into this body of water. As an example, several rivers, including the Mississippi, now flow into the Gulf of Mexico. At some time in the geologic future it is possible that the area now known as the Gulf of Mexico will be dry land and the center of the United States will be a Gulf. Now before anyone starts evacuating the center of the United States or starts planning a golf course in the middle of the Gulf of Mexico, I must warn not to be hasty because this will not happen in our lifetime, nor in many lifetimes to come, if it ever does happen at all.

These ancient bodies of water will be referred to many more times throughout this book. They may have been seas, gulfs, or embayments but they are now referred to as "basins," so from now on, I'll use the term "basin" interchangeably with gulf, sea or embayment.

In the past, as now, when a river flowed into a body of water, the coarser sediments such as sand settled to the bottom first and nearer the mouth of the river. The lighter sediments such as mud and silt were carried farther out into the body of water and deposited in deeper water. In warm climates lime is another type of sediment deposited. As the lime hardens into a rock, it is known as limestone. Limestone also comes from what was once a coral reef found in warm shallow tropical water. Limestone, along with sandstone, is a

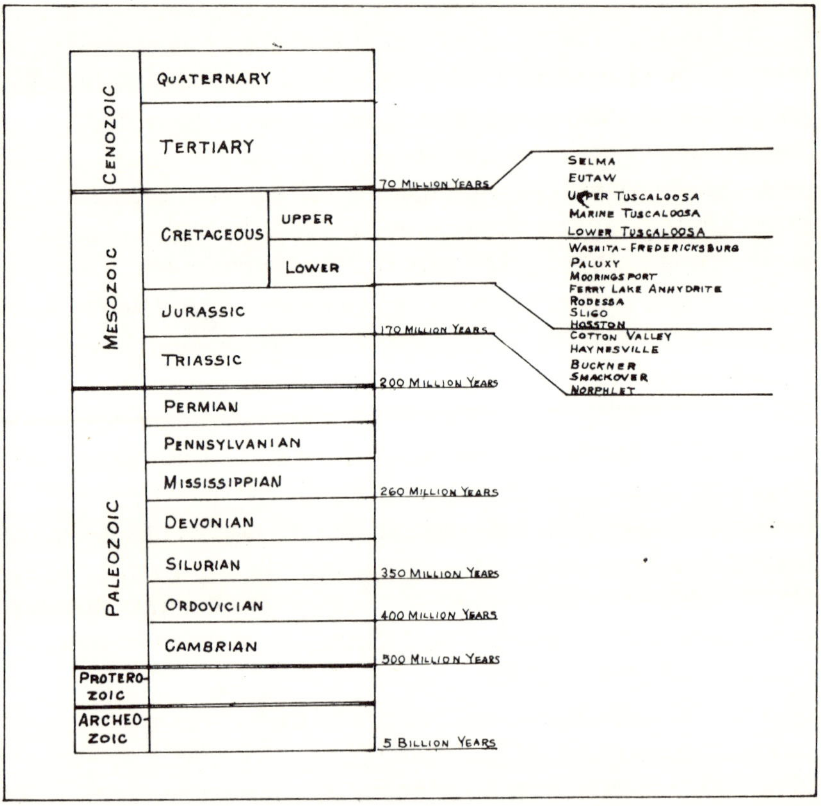

GEOLOGIC TIME CHART

Fig. 1—Any time interval in the geologic past is covered by this chart, from the beginning of the earth until the present. Each Time Period is broken down into smaller and smaller divisions until each individual sand is named if it is oil or gas bearing. As an example, since the Cretaceous and the Jurassic Periods are two very important oil producing time intervals throughout the world, I have broken them down into smaller divisions that might be familiar to you if you live in an oil producing area.

very important rock for producing oil and gas. In fact, a limestone reef is usually better than a sand because of the larger pore spaces present.

Millions of years ago the oceans were, as they are now, teeming with life. We are aware of much of the present life because of the many fish that are caught or seen, etc. . .but there is also a tremendous amount of life present that we are not aware of. Some life forms can be seen only with a microscope; some are larger but still so small that they can barely be seen with the naked eye. All of this life that is not eaten by larger fish dies and sinks to the bottom of the sea. When these organisms sink to the bottom, they accumulate and are mixed with the mud and silt. This process occurred millions of years ago also. It is important to realize that tons of these small organisms died and sank to the bottom. Eventually, there were hundreds of feet of this mud with the organisms all mixed in. As I stated earlier, forces deep within the earth are constantly at work causing certain parts of the surface to rise and certain parts to sink, over a period of millions of years.

As the land rises, the sand from the rivers is deposited farther from the original shore and is deposited on top of the mud and organisms. Then after millions of years pass, the land starts to sink again and the seashore gets back to its original position, thus depositing mud again on top of the sand. As this sequence happens again and again, eventually there will be alternating layers of mud, silt, sand and lime present. (Figures 2 and 3)

The tremendous weight of these sediments will cause the mud, which is softer, to be compressed into only a fraction of it's original thickness. Chemical reactions of bacteria, pressure, and temperature will cause the organisms to change into oil. You might say oil is formed from the juices of these small "bugs." This is a theory of course since no one was around to witness the process, but it is the one that is most accepted. It is known as the Organic Theory.

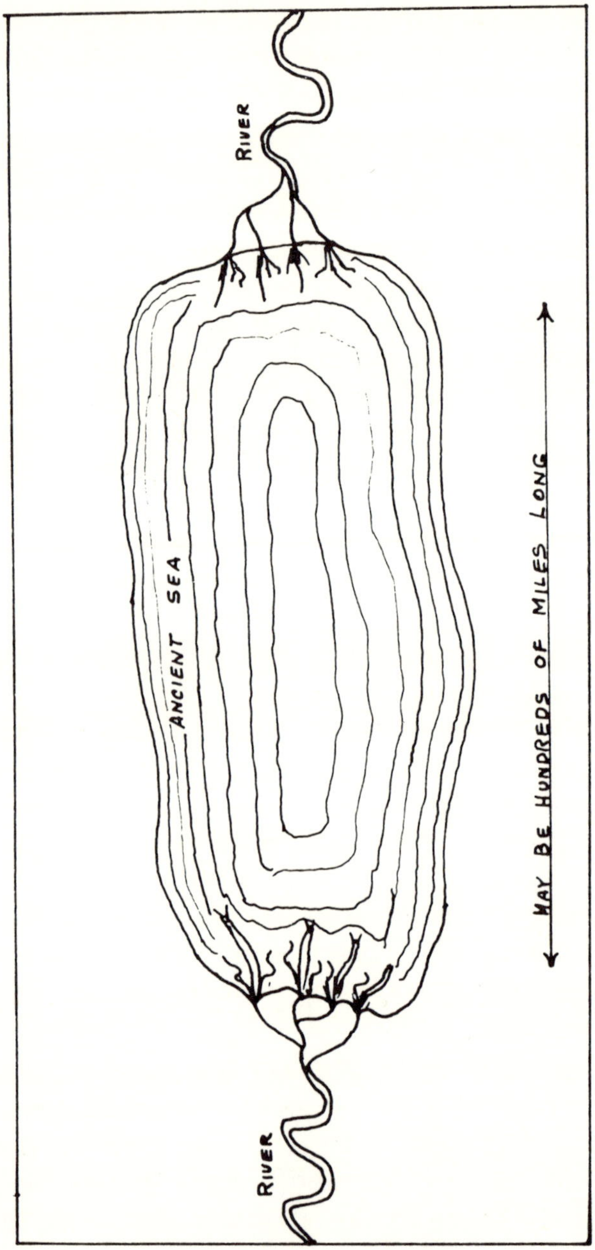

THE BEGINNING OF OIL

Fig. 2—This is how the surface of some parts of the United States looked millions of years ago. Tons of sediments were deposited over a period of thousands of years by the rivers flowing into the ancient seas.

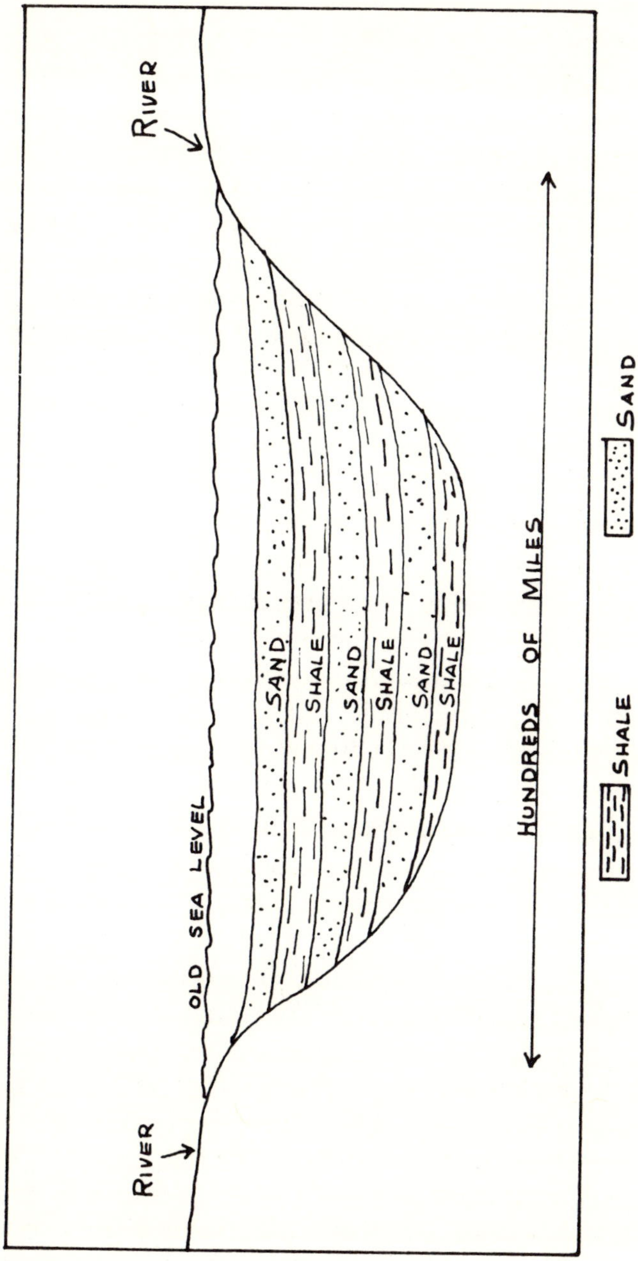

Fig. 3—Side view of Figure 2, showing how the sand and shale sediments were deposited in alternating layers. Sediments of this type can be as much as 30,000 feet thick or more.

Someone has estimated that it takes 200 feet of solid "bugs" compressed together to make one inch of oil.

As more and more sediments are dumped on top of the existing sediments, the mud becomes more and more compressed and will turn into a rock known as shale. The sand turns into sandstone, the silt turns into siltstone, and the lime turns into limestone. As the mud becomes compressed into shale, the oil is squeezed out of the shale and into the pore spaces between the sand grains of the sandstone or the pore spaces in the limestone. The larger pore spaces of the limestone and sandstone will accommodate the oil more readily. (Figures 4 and 5) Therefore, the mud or shale has been the source of the oil and the sandstone or limestone acts as the reservoir to hold the oil.

Now that I have mentioned the main types of sediments that are connected with oil production—sand, silt, shale and lime—it will be much easier reading if I just refer to sand and shale from this point on. As a reservoir rock, sand and lime are basically the same; each time I mention sand or sandstone, the same information generally is true of limestone.

To avoid confusion, it would be a good idea to fully explain what is meant by a reservoir, because the term will be used several times throughout the book. A reservoir is merely a solid sand formation in which oil has accumulated. If you covered a sand beach with tons of sediments, you would have a potential sand reservoir. The only difference is that the loose sand grains would consolidate into a hard sandstone rock. If you have always imagined that oil is in a huge cavity beneath the surface, this might be a difficult concept to accept. The oil is drawn from between the sand grains to the well and then flows or is pumped up the well into storage tanks. This sounds as though it is a simple procedure but it is far from being simple. Great care must be taken to keep from damaging the formation around the well bore, thus ruining the well.

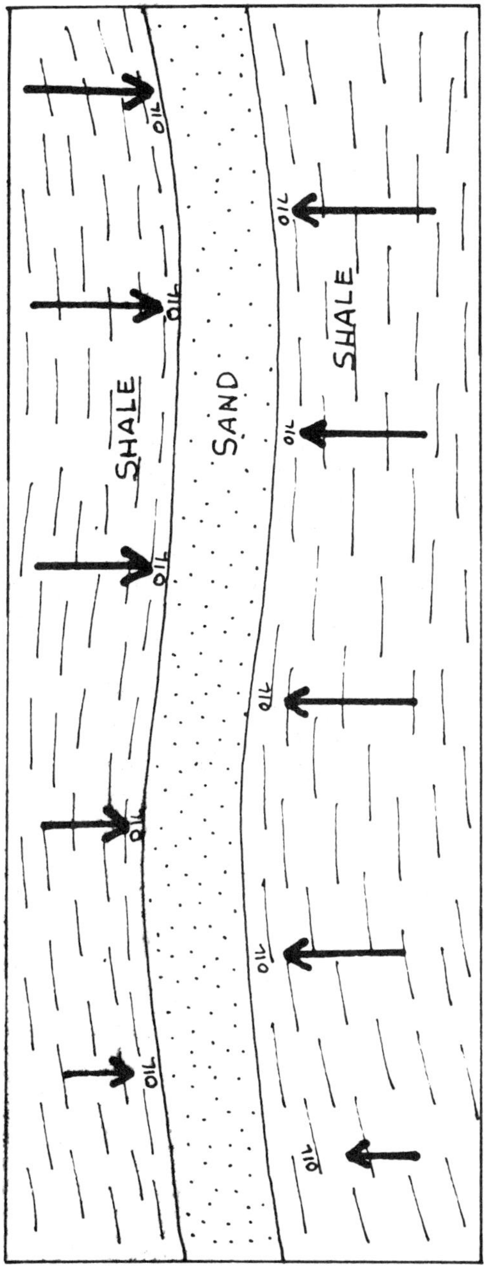

OIL IS SQUEEZED INTO THE SAND

Fig. 4—Because of the tremendous weight of the sediments above, the oil is squeezed out of the shale and into the sand.

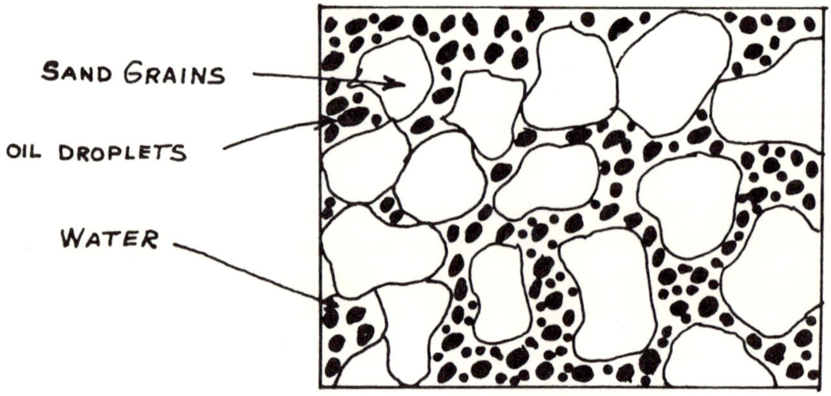

Fig. 5—A microscopic view of an oil bearing sand formation, showing how oil and water are found in between the individual sand grains. You can see by this illustration that oil is not found in large caves or cavities.

Birds migrate . . . so does oil

Now that the oil is in the sand, it is free to move about or "migrate." As the sediments and organisms were being deposited, salt water was also deposited and trapped in the sediments. Since oil is lighter than water and will not stay mixed, the two fluids will separate and the oil will be on top of the water. In separating, the oil will appear to flow up hill, but it is merely seeking it's way to the top of the salt water. (Fig. 6)

Why do some wells flow naturally?

Very often, as the oil is being formed, gas is also formed in the process. Sometimes the gas will stay mixed with the oil and sometimes it will separate, depending on how much gas is present. The gas that is mixed in with the oil is what causes the pressure enabling some wells to flow naturally. If there is no gas present in the reservoir, a well must be pumped.

The oil and gas move through the sand trying to get above the water until it reaches a barrier and can go no farther. When it reaches this barrier, it is said to be trapped. These "traps" are what we look for in our search for oil. (Figures 7-12).

Earthquakes . . . what causes them?

There are constant movements of rock formations deep beneath the surface of the earth. Usually the formations break and slide over a period of many years and the movement is so gradual that it is not detected at the surface. Ever so often the movement is rapid enough to be detected, but not near enough the surface to do any damage. If the fracture does extend to the surface, we have a major earthquake. (Figure 12)

Now that the oil has formed and migrated to a trap, it will stay there for millions of years waiting for man to tap the vast energy that will lay dormant for so long.

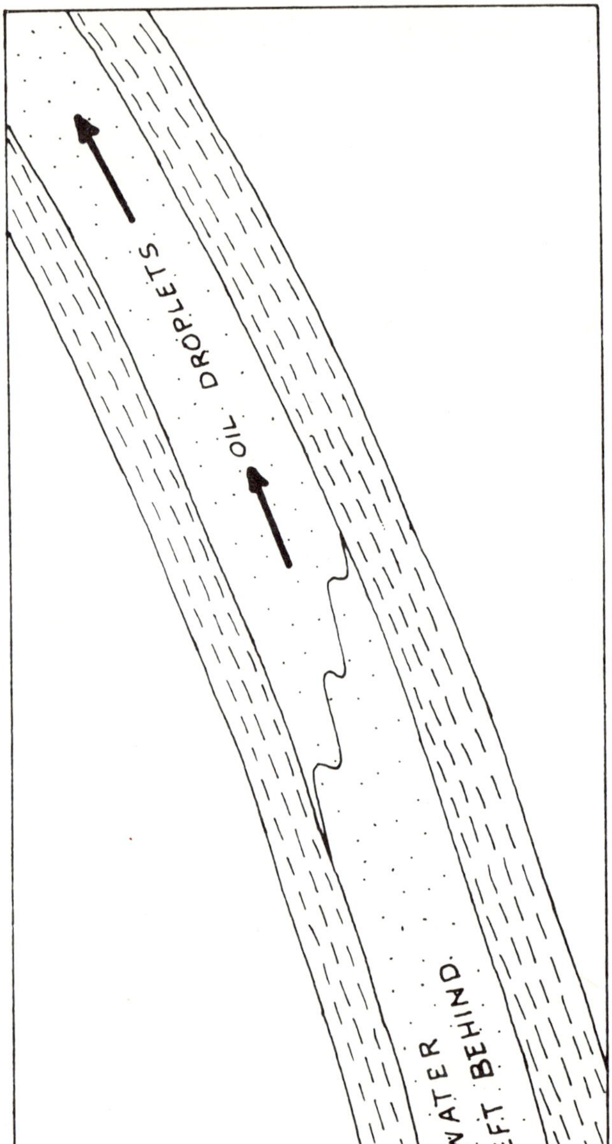

MIGRATION OF OIL

Fig. 6—Since oil is lighter than water, the oil droplets will move (float) upward and be above the water. The oil and water cannot get out of the sand because the shale acts as a seal above and below the sand. The oil will continue to move upward within the sand until it reaches a barrier and can move no farther.

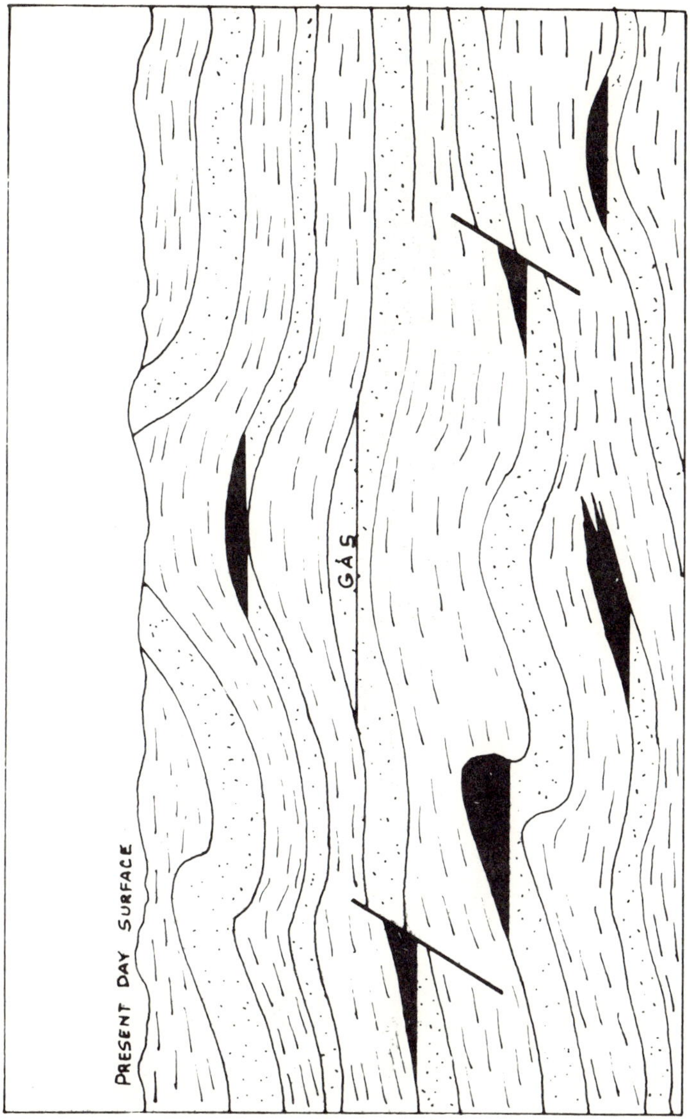

Fig. 7—After the sediments are deposited, tremendous forces deep within the earth cause the formations to be twisted, arched, and fractured, causing many types of traps for oil to accumulate. These same forces have pushed these sediments upward for thousands of feet. As this happens, the surface of the ground is eroded away, so that sediments that were once thousands of feet below the surface may now only be a few hundred feet below the surface.

This illustration is just meant to show the various types of traps. Please do not think that oil is this pentiful beneath the surface.

STRATIGRAPHIC TRAP

Fig. 8—A very porous sand may abruptly turn into shale or an impermeable silt. When this happens and the oil can move no farther, it is trapped by a change in stratigraphy (a change in the composition of rocks). Some of the largest oil fields are stratigraphic traps.

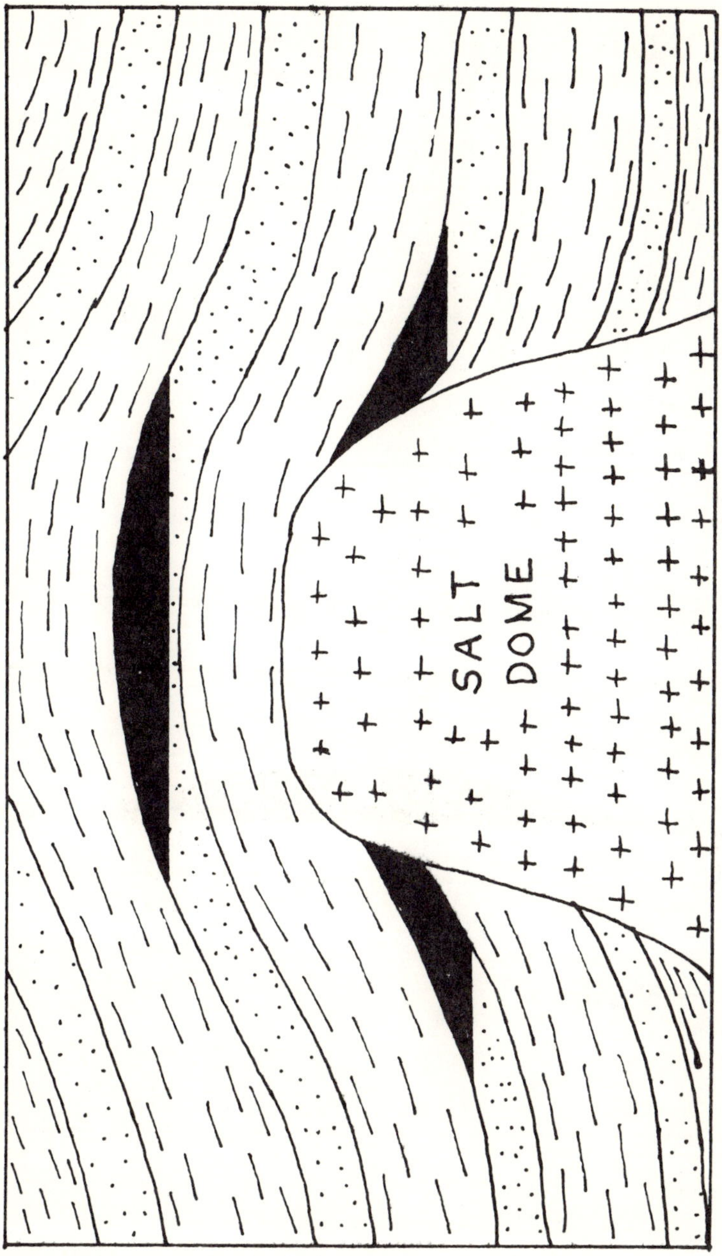

SALT DOMES

Fig. 9—This type of trap is discussed later in a chapter on salt domes.

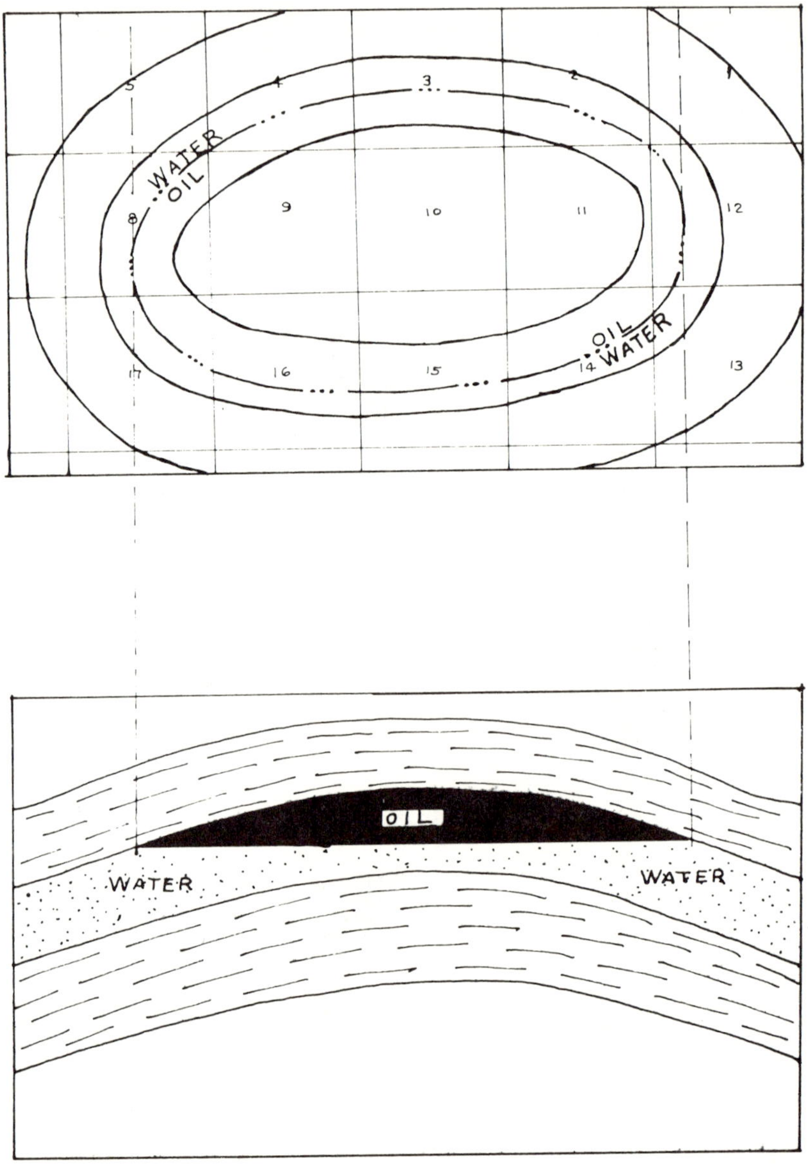

ANTICLINE

Fig. 10—Top and side view of an anticline, which is simply a "hill" under the ground. The oil or gas accumulates at the top of the "hill." This is the most common type of trap.

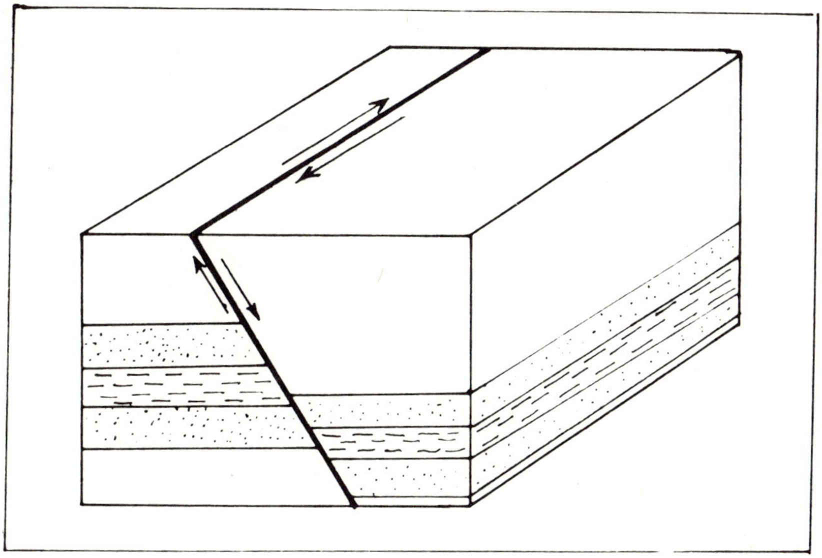

Fig. 11—A block diagram showing that as the formations break and move beneath the surface, they move in a lateral direction as well as a vertical direction.

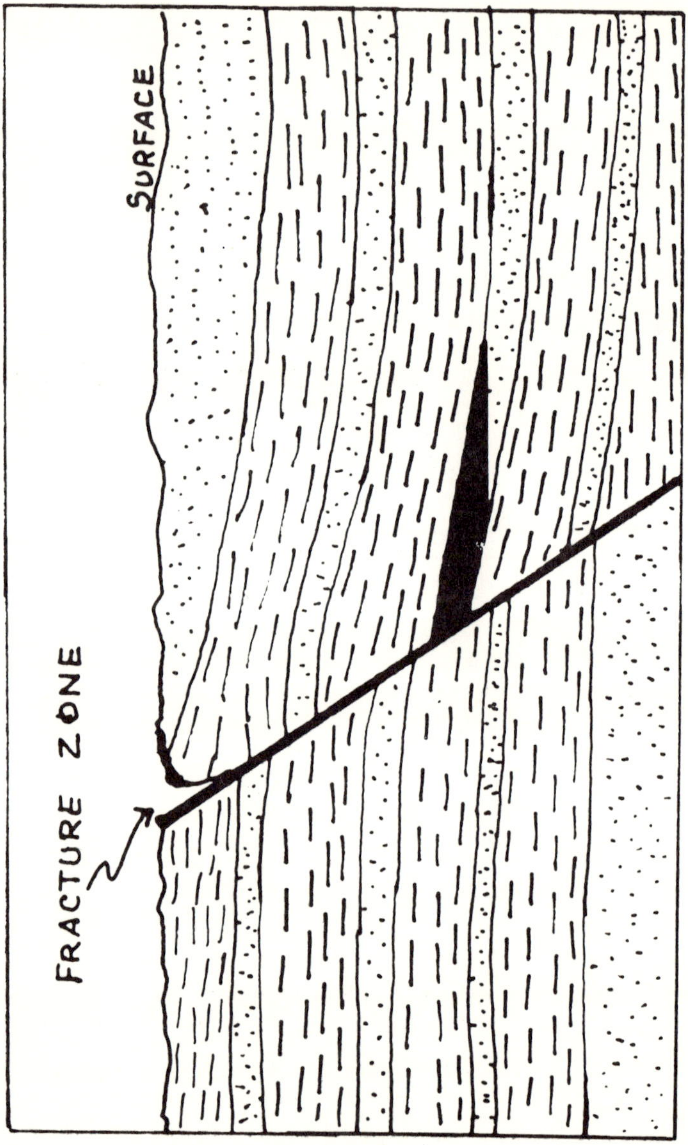

FAULT TRAP

Fig. 12—As the formations break and are offset beneath the surface, oil will sometimes accumulate against the break, which is called a fault.

If the beds move slowly and do not break all the way to the surface, it is called a fault. If the movement is rapid and does extend to the surface, it is called an earthquake. This figure shows a fault extending all the way to the surface, causing the surface to fracture.

Chapter 4

The search begins

Millions of years have passed and man has finally appeared on the scene. He probably first found oil accidentally in a tar or an oil seep on the surface of the ground. The tar was possibly used to make his boats watertight and he burned the oil to generate heat and light. As man came to realize how important oil was to his life, he knew that he could not rely forever on oil seeps to supply his total need. So he started searching for oil.

Man searched for oil in many ways and in many places until finally the oil industry was born in the United States in 1859. The first commercial oil well was drilled by "Colonel" E. L. Drake in Titusville, Pennsylvania. It was drilled to a depth of 69½ feet. By comparison, the deepest well in the world was drilled in 1970 off the coast of Louisiana in the Gulf of Mexico. It was drilled to a depth of 25,600 feet. Even now, there is a well drilling in Texas which has an objective depth of 27,500 feet. If it is successful in reaching that depth, it will establish still another world record.

Most of the first wells were drilled using surface indications, such as oil seeps; many were drilled solely on a hunch; but any oil that was found was largely a matter of luck. The story goes that in the early days of drilling for oil, an individual had a small drilling derrick mounted on the back of a truck. He was on his way to drill at a location that he had picked out, but several miles short of his destination the truck broke down. It so happened that the truck was not the only thing that was "broke" . . . the driller was too. So broke in fact that he could not pay to have the truck repaired so he just drilled a well where he was . . . and found one of the largest oil fields in Texas. Later he drilled at the original location that he had picked out and drilled a dry hole.

Another driller was supposed to drill a well in the center of a certain Section in a certain Township. He drilled in the center of the Section but in the wrong Township, 12 miles from where he was supposed to be. Result? . . . he also found a new oil field at the wrong location. The odds of finding oil by this method are astronomical but nevertheless, it has happened on rare occasions.

People have used electrical boxes, divining rods and every other method conceivable to try to detect oil beneath the surface of the ground. But the only way to know for sure whether oil is beneath the surface is to drill a well.

How do we decide where to look for oil?

First we go to a place that is favorable for finding oil. In order to do this, geologists have to reconstruct the geologic past and determine where the ancient seas that existed millions of years ago are located beneath the surface of the earth today. Geologists first examine and study the rocks that are present on the surface. The most favorable areas in which to study the rocks are along river banks and fresh road cuts where the rock formations are laid bare. Of course there are many areas in our country where rock formations are found naturally exposed at the surface, such as in a very rocky area where there is not much vegetation.

These various formations are then studied to see in which direction and how steeply they dip or slant beneath the surface. By projecting these formations down to thousands of feet beneath the surface, the boundaries of these ancient seas are found. This explanation is over-simplified of course; the actual study is much more complex.

We not only have to find the ancient seas, but we also have to find the sand and shale present in the correct ratio to each other. The correct amount of shale has to be present for the oil to have formed and the correct amount of sand has to be present to serve as a reservoir for the oil. We can determine the presence of the ancient seas and we can determine if the correct ratio of sand and shale is present but more detailed study is necessary in order to pinpoint the potential oil traps.

The ancient sea that we have discovered may cover several hundred square miles. Within these limits there are possibly dozens or even hundreds of oil fields in all shapes and sizes. The problem, of course, is finding these fields. The cost of drilling a well varies from 25 thousand dollars to over two million dollars depending on several factors, such as depth drilled, hardness of the rock formations and other technical problems that will be discussed in the chapter on **A Well Is Drilled.** Because of the high cost of drilling a well, a tremendous amount of planning and testing has to be done in order to cut down the odds of drilling a dry hole as much as possible. Even then, utilizing the most scientific methods available, approximately eight out of every nine wells drilled are dry holes.

Out of the hundreds of square miles that can be a potential oil producing area, how do we decide exactly where to drill a well? There are four basic ways that we decide:

1. Surface geology

Just as the name implies, this is a method by which the surface of the earth is studied and predictions are made as to what type of structures lie beneath the surface. This is the oldest scientific method and the most frequent method used in the early days of oil exploration. It must be kept in mind that a geologist can never tell if there is any oil present below. What he looks for is a structural trap that might contain oil. Sometimes the structure that exists thousands of feet beneath the surface is reflected at the surface and can be seen. (Figure 13) Most of the time however, a hill or mound present at the surface is due to erosion and does not mean there is a similar structure deep below. (Figure 14) This determination can easily be made by a geologist.

2. Gravity meter and magnetometer

These two instruments measure the density and magnetic properties of the rock formations beneath the surface of the earth. These instruments are utilized by flying over an area in an airplane. Only large structures can be found by this method; smaller structures do not usually show up on the records.

3. Seismograph

More oil has been found by the seismograph or seismic method than by any other means. This is the method by which small charges of dynamite are exploded in a hole that is drilled about 75 feet deep. The sound or shock waves go down into the earth and as they hit the different layers of sand and shale, they bounce back up to the surface and are recorded by small timing devices that have been placed on the ground. By measuring the amount of time that it takes for the sound waves to bounce off a formation and come back to the surface, it can be determined how deep the formation is below the surface of the ground. (Figure 15)

I would like to correct another misconception at this point. Some people have thought that when these shallow 75-foot holes were drilled, the oil company was actually drilling for oil. This is a false impression; the only reason for drilling the shallow holes was to explode the dynamite. It has also been

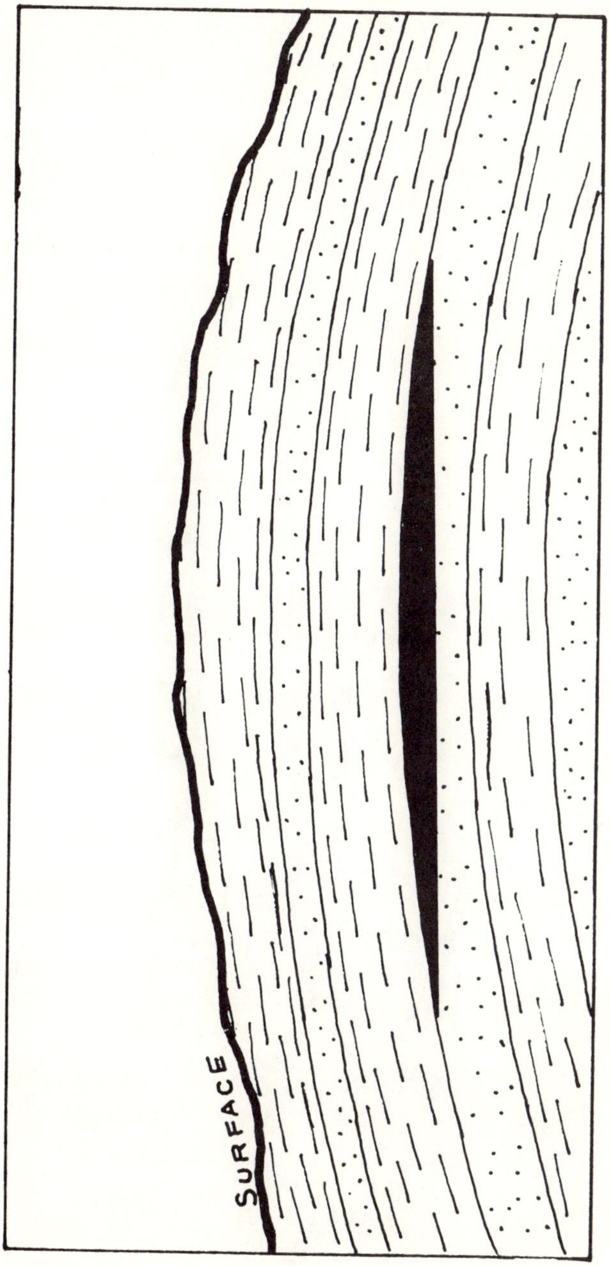

Fig. 13—If there has not been much erosion at the surface, the structure or contour of the surface is the same as the formations below the surface. In this case, a geologist would recommend drilling on top of a hill at the surface.

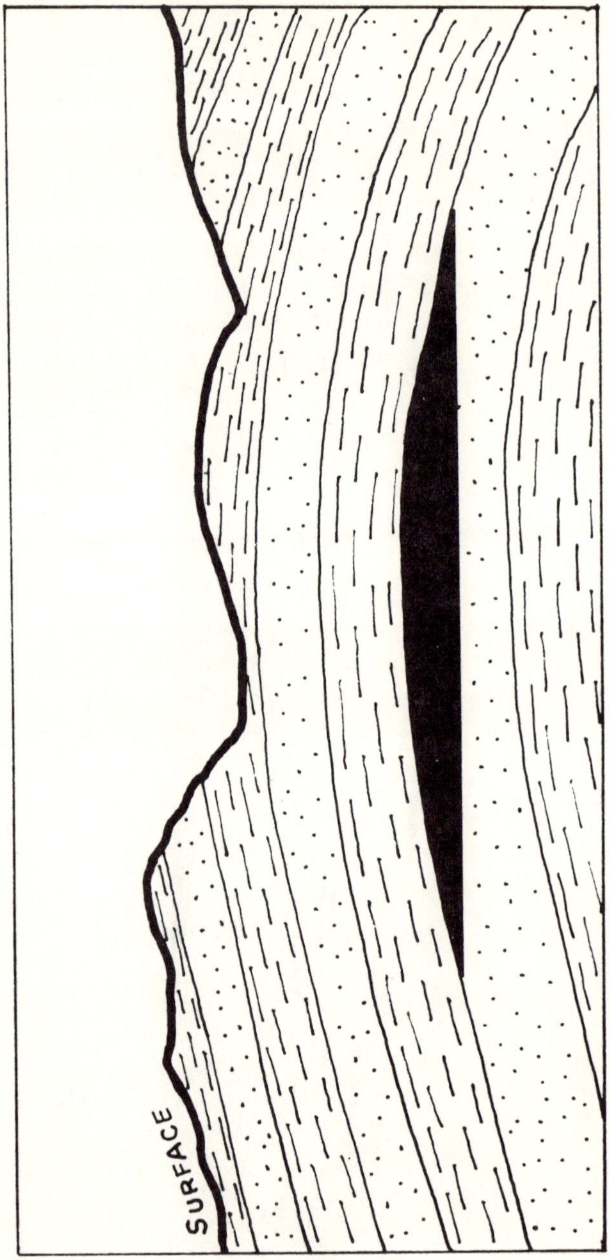

Fig. 14—In most cases, the surface has been eroded by wind and water and does not reflect the structure of the formations beneath the surface.

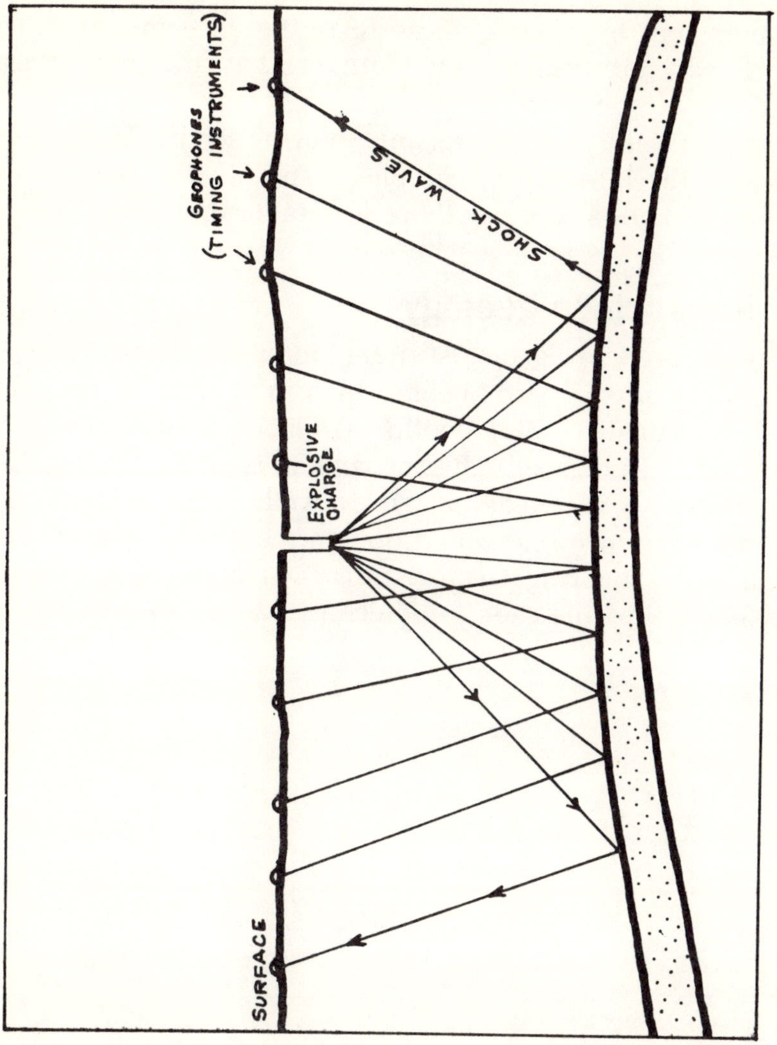

SEISMIC METHOD OF PROSPECTING FOR OIL

Fig. 15—A hole is drilled to a depth of about 75 feet, in which a small charge of dynamite is exploded. Prior to the explosion, several timing devices are placed on the ground in a straight line on both sides of the hole. These are very sensitive instruments and measure the speed of sound in fractions of a second. As the dynamite is exploded, the sound or shock waves go downward into the earth. The shock waves hit the formations, bounce back to the surface and are recorded by the timing instruments. By knowing how fast the sound travels and measuring how long it takes to bounce off a formation and back to the surface, the depth and contour of the formation can be calculated.

thought that by making these tests the oil company knew there was oil below. The truth is that many of these shallow holes are drilled systematically over several square miles. It is only after the seismic information has been taken back to the laboratory, compiled, and studied that an oil company knows whether there is a favorable place to drill anywhere in the area just tested.

4. Subsurface geology

By this method a geologist studies information gained from wells that have already been drilled in an area. After each well is drilled to total depth, an electrical instrument is lowered down the well. This instrument is an electric log and the process is known as "logging a well." The log shows the depth of the formations and, to some degree, if any of the sands are oil bearing. Geologists use this information to draw geologic maps in order to determine where to drill future wells.

Chapter 5

Preparing to drill a well

Before an oil company can drill a well, it has to buy an oil lease on the land to be drilled. Leasing is discussed in detail in a later chapter, so to simplify the situation, we'll just assume that the land is leased.

Why does my neighbor get part of the oil if the well is drilled on my land?

Before a company starts to drill the well, it has to designate a drilling unit around the well consisting of a certain number of acres. The law varies, but most states allow only one well to be drilled on every 40 acres. If the well is drilled in an area where there are a number of small tracts of land, several landowners could possibly own acreage within the drilling unit. Just because a well is drilled on a certain tract of land, it does not mean that the owner of that particular tract will receive a larger share of the oil production. It all depends on the number of acres an individual owns within the drilling unit. (Figure 16)

An oil company also has to apply to the state for a permit to drill a well. In the application for the permit a company must submit a plat showing the location of the well and the drilling unit around the well. The well location and the drilling unit must be surveyed by a registered engineer. The company must also state how deep it plans to drill the well. If all is in order, the state will issue a permit to drill.

Could you drill a well in your back yard if you wanted to?

You could not drill a well in your back yard unless your back yard contained at least 40 acres. If you owned only 10 acres you would have to get your neighbors to join you with another 30 acres in order to form a 40-acre drilling unit. Even then, you would have to apply for a permit to drill a well, just as an oil company has to do. (Unit size varies by states.)

You may have seen pictures of oil fields in Texas or California where the derricks were practically touching each other. These wells were drilled many years ago before there were regulations; some of these wells were drilled in back yards that were nothing more than city lots. Each person had to drill a well on his lot for self protection. If he didn't have a producing well on his land he didn't get any oil payments.

The reason for the present regulations is simple: if one well will adequately drain 40 acres, why drill 10 wells to drain 40 acres?

Drilling units for gas wells are larger than drilling units for oil wells. The reason is that since gas flows through the reservoir much more freely than oil, a gas well will drain a much larger area than an oil well will drain; therefore, fewer wells are drilled in a gas field than in an oil field of comparable size.

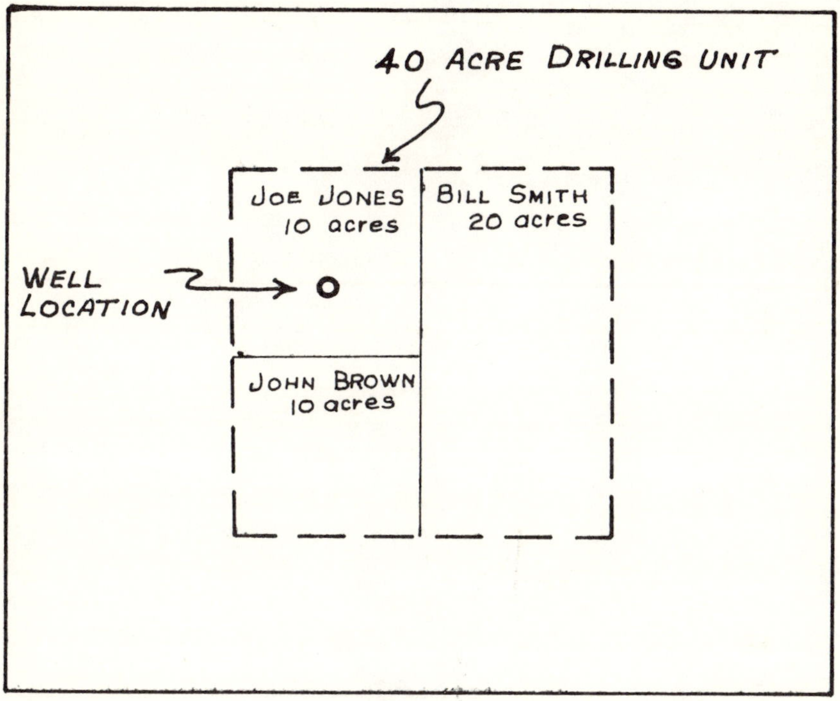

DRILLING UNIT

Fig. 16—Suppose a well is drilled on Joe Jones' 10 acres. He has 10 acres of the 40-acre drilling unit, so he would receive ¼ of the royalty payments from the well—so would John Brown, who also owns 10 acres in the unit. Even though the well is not drilled on Bill Smith's land, he owns 20 acres within the unit; thus he would receive ½ of the royalty payments.

It does not matter where the well is drilled within the unit. Royalty payments are based on how much land one has within the unit.

Just as the size of oil units vary, the size of gas units also vary, but a common practice is to drill one gas well on each 320 acres. Whatever the size of the drilling units in an oil or a gas field, all wells in the field must be drilled on units of equal size.

Chapter 6

A well is drilled

Now that the leasing, surveying, testing, mapping and legal work is completed, we are ready to spud (start) the well. If a well is drilled in an area where there are no producing wells it is called a "wildcat." If the "wildcat" well finds a new oil field, all wells drilled in the field are called "field wells." The drilling that will be discussed within the next few pages will be considered a "wildcat."

First, of course, the location has to be cleared and leveled so that the derrick will be standing vertically. Depending upon the size of the derrick, an acre or two is usually required.

Methods of drilling a well

There are two methods used in drilling a well: the cable-tool and the rotary method. The cable-tool is a method which was used in the earliest days of drilling and is still used on many shallow wells. The drilling bit is supported by a cable at the surface of the ground and the hole is drilled by a peck and chip action. The bit is lifted and dropped several times a minute, thus drilling the hole.

By far the most popular and versatile method used today is the rotary drilling method. The drill bit is connected to lengths of hollow drill pipe and as the drill pipe is rotated at the surface by heavy and powerful machinery, the drill bit rotates and grinds out the hole.

$50,000.00 for mud?

Strange as it may seem, this amount could very easily be spent in drilling a normal well to a depth of 12,000 feet or more. If the well has certain technical problems, the mud cost could run into hundreds of thousands of dollars. Of course this is not the kind of mud you made pies with when you were a child, but nevertheless, it is mud and it is called mud. This is a chemical mud and the well could not be drilled without it. There are several reasons for this, which are discussed a little later in this chapter. The mud is mixed in a shallow pit that is located next to the well; then it is pumped out of the pit through a pipe up to the derrick floor, down the center of the drill pipe, out through the drill bit, up the hole on the outside of the drill pipe to the surface and back into the mud pit. The mud makes this complete cycle thousands of times during the drilling of a well. The drill bit is larger than the diameter of the drill pipe so that the mud can return to the surface between the drill pipe and the sides of the hole. (Figure 17)

Is this expensive mud really necessary?

There are several vital reasons for having drilling mud:

1. Helps prevent "blowouts" (gushers) and well fires.

2. It carries the pieces of rock that the drill bit grinds up back to the surface.

3. It keeps the drill bit cool; the temperature at the bottom

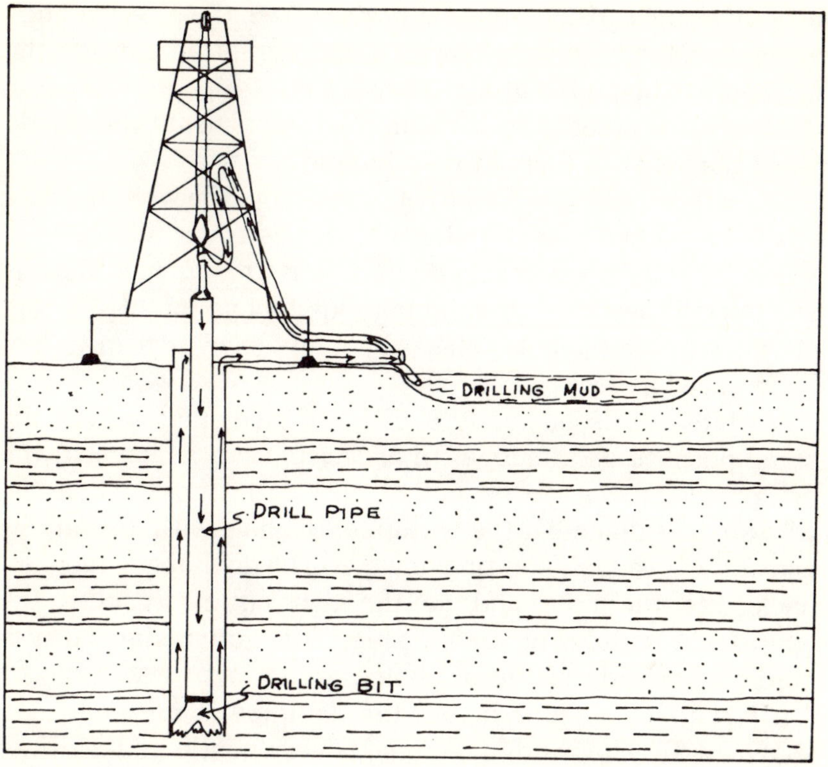

ROTARY DRILLING RIG

Fig. 17—This type of drilling got it's name because the hole is drilled by rotating the drill pipe. The pipe is rotated in a clockwise direction at the surface causing the drilling bit to rotate and drill the hole. Drilling mud is continuously circulated down the inside of the drill pipe and back to the surface. The drilling mud performs several functions and is vital to this type of drilling.

Drill pipe comes in 30-foot joints. When a drilling bit wears out and has to be replaced, the pipe is pulled out of the hole. The joints of pipe are unscrewed and stacked against the side of the derrick. This is called ''making a trip.''

of the well can get higher than 400 degrees F. in a deep well.

4. It forms a wall cake or temporary seal on the sides of the hole and helps keep the sides from caving in.

Sometimes the walls do cave in, however, and the drill pipe becomes stuck. If it cannot be worked free, the pipe is pulled at the surface with very powerful motors, often causing the drill pipe to be pulled apart. The section of pipe that is lost in the hole is referred to as a "fish." Attempts to get the "fish" out of the hole is referred to as "fishing."

When the "fish" cannot be retrieved from the well, the hole has to be junked and abandoned or it has to be sidetracked. When a well is sidetracked, it means that a new hole is started out the side of the old hole just above the "fish" that has been left in the hole. This is accomplished by using a tool called a whipstock.

A whipstock is basically a piece of steel that serves the same purpose as a wedge. In order to use a whipstock, the drillpipe first has to be pulled out of the hole. Then the whipstock is placed at the bottom of the hole with the narrow end of the "wedge" pointing upward. The drillpipe is then put back into the hole and as the drilling bit reaches the whipstock, it slides off to one side. This action starts a new hole out the side of the old hole.

Contrary to popular belief, the drillpipe is not bent at a sharp angle. The bend in the pipe is very gentle and does not cause the strain that a sharp bend would cause.

When a well is drilled off to one side in this manner, it is called a "directional well." There are several reasons for drilling directional wells; they will be discussed later.

5. After a well is drilled, the mud acts as an electrical conductor for the instruments used in evaluating the various formations.

The chemical composition of the mud is very important during this stage of evaluating the well. The instruments are calibrated with the chemicals in the mud and an accurate measurement of the chemicals is vital in order to record correct readings.

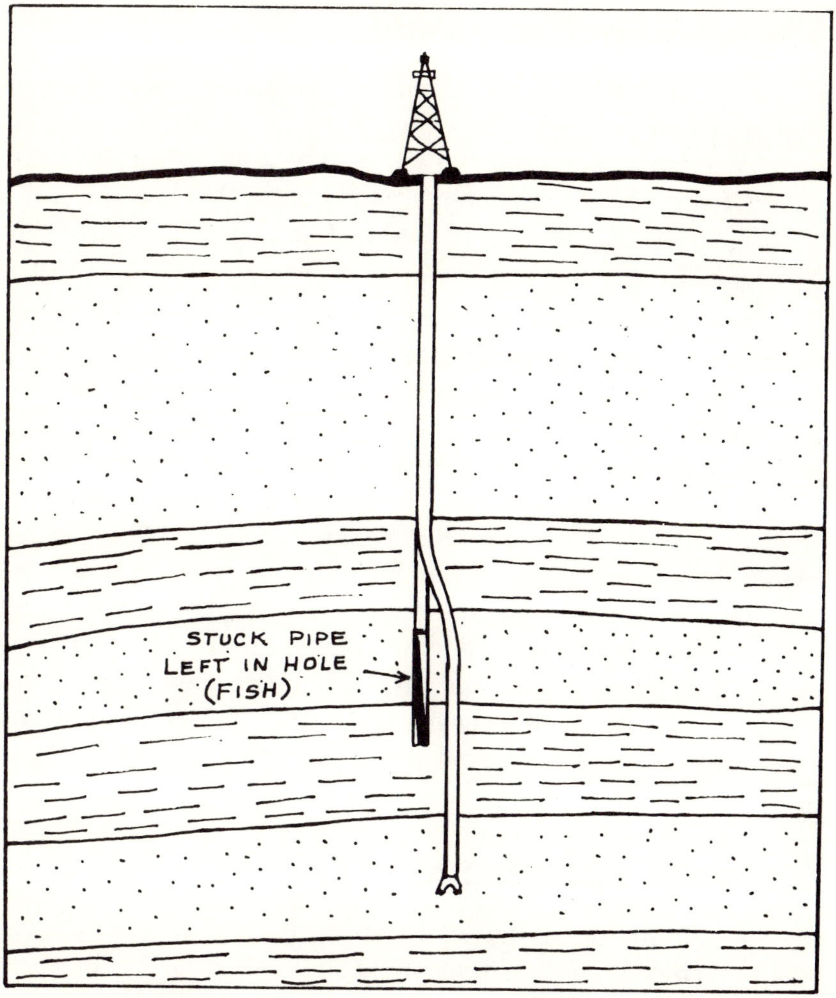

Fig. 18—The walls of a well cave in sometimes and causes the drill pipe to become stuck. In trying to work the pipe loose, it is often pulled in two and part of the drill pipe is left in the hole. The portion of pipe left in the hole is called a "fish" and the process of trying to get the "fish" out is known as "fishing." If the attempt fails, a new hole just above the "fish" is started and the well is drilled around the pipe left in the hole. A tool that enables a well to be drilled at an angle is called a "whipstock."

Are fresh water sands damaged when drilling a well?

During the first few hundred feet of drilling, fresh water sands are encountered. The deeper sands in the well are all salt water sands, so something has to be done to keep the salt water from coming up the hole and contaminating the fresh water sands that are most likely being used for drinking water.

It is a simple matter to do this. After all the fresh water sands are drilled, the drill pipe is pulled out of the hole and a larger permanent pipe called "casing" is put into the hole. Cement is pumped down the inside of the casing and back up the hole on the outside of the casing until it reaches the surface. The exact amount of cement that is needed is calculated by an engineer beforehand and a plug is pumped down the inside of the casing on top of the cement. When the plug gets to the bottom of the hole, the engineer knows that the casing is enveloped in cement from top to bottom. This is a special cement that hardens in just a few hours. The cement has sealed off every fresh water sand and the casing is permanently in place so drilling can now continue. (Figure 19)

Drilling continues

The remainder of the well will be drilled with a smaller drill bit since it now has to pass through the casing in order to reach the deeper formations that are left to be drilled. Unless a problem occurs, the well will continue to be drilled 24 hours per day until the total depth of the well is reached.

The men who drill the well are referred to as a drilling crew and usually consists of four men. One is the "driller" and the other three are "roughnecks." These men have to know what they're doing and be alert at all times. They often have to react very quickly to keep a minor problem from

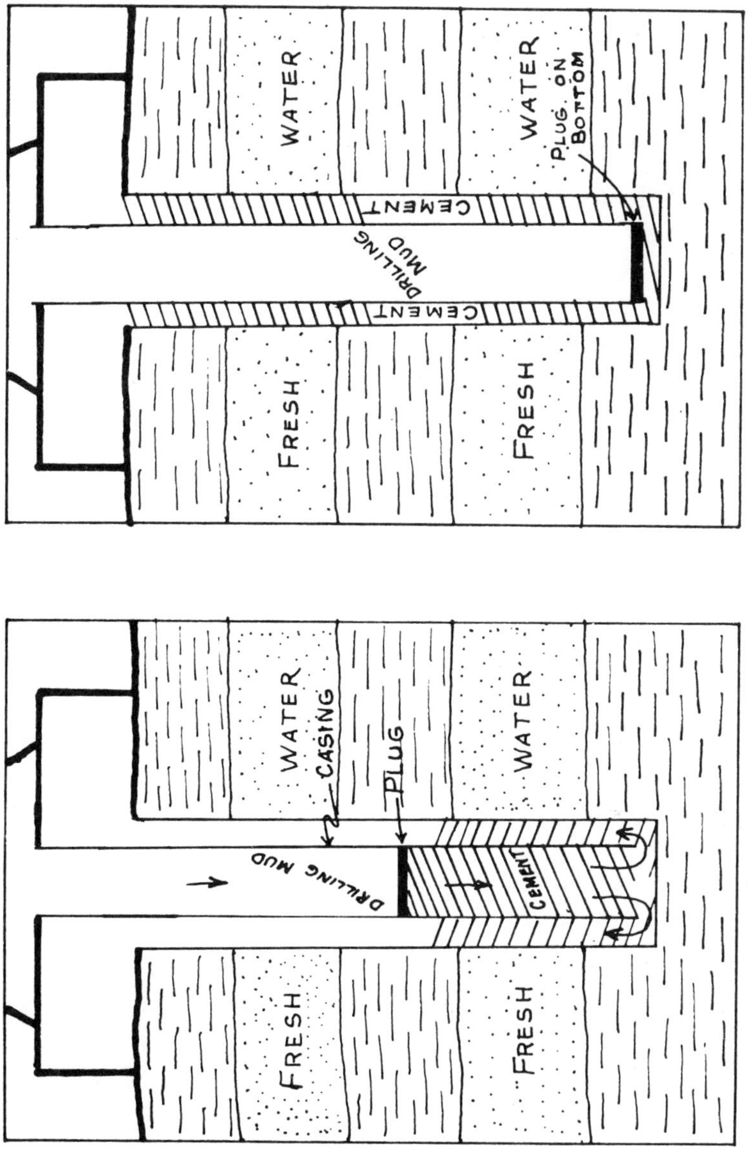

Figure B

Figure A

FRESH WATER SANDS ARE PROTECTED

Fig. 19—Cement is forced down the inside of the casing and up the outside between the casing and the walls of the hole. When cement is in place it will harden, thus sealing off the fresh water formations.

becoming a very costly problem, as evidenced in the next paragraph.

What causes a "Gusher"?

As the oil is formed and migrates into a trap, there is usually some gas that is associated with the oil. This means that the oil could possibly be under very great pressure, depending upon the amount of gas present. The term "gusher" came into use in the early days of drilling for oil. Since it makes an exciting scene, practically all motion pictures about oil show a well "blowing wild" or "gushing" oil up out of the ground. The owner of the well who is always broke and about to lose everything he owns is gleefully dancing about and hugging everyone in sight as the oil is falling like rain, covering everything and everybody.

This does make an exciting movie but you can be sure that if that happens today, everyone working at the well is doing everything within his power to stop the flow of oil. Not only is this a waste of oil and money, but there is also extensive damage to the surrounding surface of the ground for which the oil company is responsible. When a "blowout" does occur today it is usually when a high pressure gas formation rather than an oil formation is drilled into.

Two things help prevent "blowouts": the drilling mud and a large valve called a "blowout preventer." By adding certain chemicals, the mud can be made heavier or lighter, depending upon the need. The weight of the mud can vary from nine pounds per gallon to over 19 pounds per gallon. By comparison, water weighs approximately eight pounds per gallon. In order for oil or gas to "gush" up out of the well, it would first have to push thousands of pounds of mud out of the hole. It is usually anticipated how much mud weight is needed to overcome the pressures that might be encountered while drilling, and the mud is prepared accordingly.

Even when preparations are made, however, unexpected high pressure zones are sometimes drilled into and the pressure is powerful enough to blow even the heavy mud out of the hole. If this happens, and the mud starts to blow out of the well, the second safeguard is used . . . the "blowout preventers." This is merely a large valve at the surface of the ground under the derrick floor. It can be closed, sealing off the well until it can be brought under control. It happens rarely, but it does happen . . . sometimes a formation is drilled into that has such a tremendous amount of pressure that it not only blows the mud out of the hole, but it also blows the derrick, along with the blowout preventers, off the foundation and usually catches fire.

When this happens, a speciality company is employed to extinguish the fire. If the casing is sticking up above the surface of the ground and not damaged too severely, the well can be capped. However, if the casing has been blown out of the well, there is no way to cap the well. The only way to "kill" the well then is to drill another well into the formation that is causing the "blowout." This second well is called a relief well. The relief well is located about one-half mile from the wild well and it is drilled at an angle beneath the surface by means of a whipstock. An attempt is made to drill the relief well into the formation at the same point that the burning well encountered the formation. Being aware of the tremendous pressure, this time the driller will make sure that the mud weight is heavy enough to contain the high pressure. When the formation is drilled into by the relief well, cement or mud is pumped down the hole and out into the formation. This procedure will seal off the burning well at its source, if the attempt is successful. (Figure 20)

Geologist studies rocks while well is drilling

On many wells a geologist will examine the bits and pieces of rocks that the mud brings to the surface as the well is being drilled. The mud passes over a wire screen before it returns

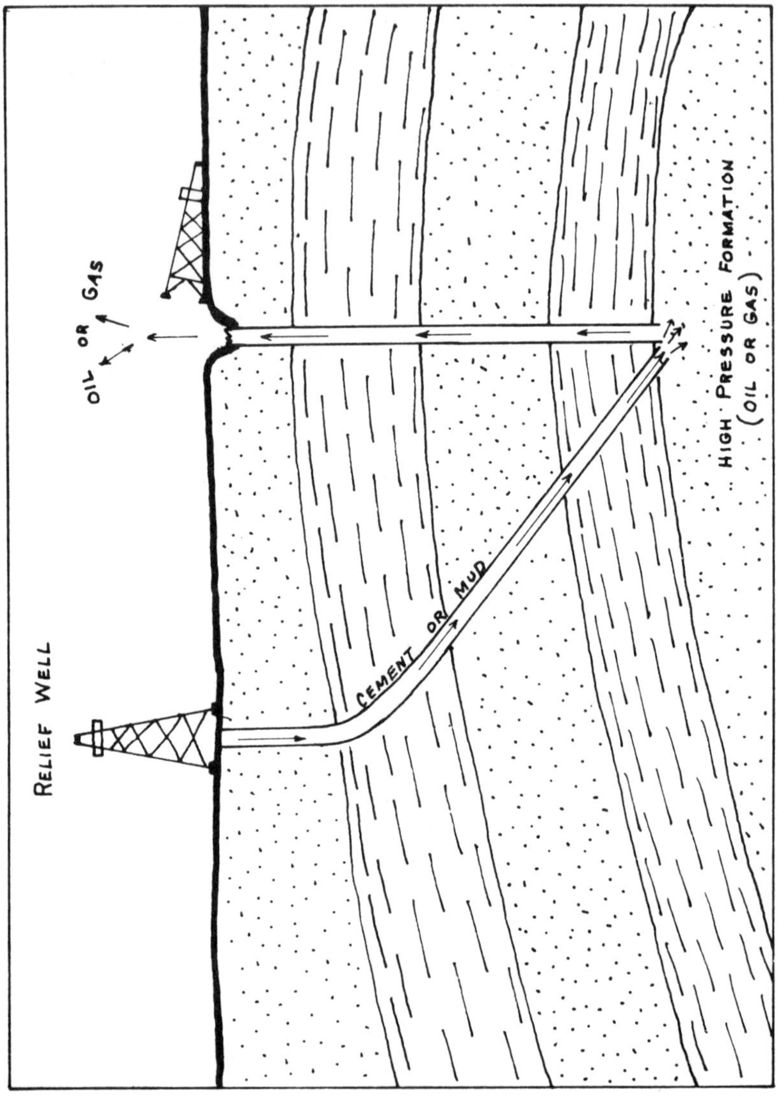

KILLING A WELL

Fig. 20—If a well "blows out" and the oil or gas cannot be brought under control by other means, a "relief well" has to be drilled. The relief well is drilled at an angle and the "wild" well is killed by pumping mud or cement into the formation from which the gas or oil is flowing. The relief well is drilled by the same method as the original well so there is usually a period of several weeks in which there is a tremendous amount of oil or gas wasted. The cost of drilling the relief well plus the loss of oil or gas can cost millions of dollars.

to the mud pit and the bits of rock or "cuttings" are collected and examined. He can see if there is any oil in the cuttings, but the amount of oil present or whether the well will produce cannot be determined at this point. This is just another piece of information that helps geologists decide if a well will produce.

As the geologist examines the cuttings, he may decide to "core the well." The drill pipe and standard drill bit are pulled out of the hole and a different type of bit is put on, This bit is hollow in the center, like a doughnut. With this type of bit a part of the formation is kept in tact inside the drill pipe instead of being ground up as the standard bit does. The section of formation that is brought to the surface is called a core and is usually about three inches in diameter and 30 feet long. A large piece of an oil bearing formation is invaluable for laboratory examination. (Figure 21)

How deep to drill

It is commonly believed that when a well is started, the oil company doesn't know how deep the well will be drilled—just drill away until you strike oil. This idea is not correct. As I mentioned earlier, when an oil company applies for a permit to drill, it must state not only where the well will be drilled, but also how deep it will be drilled. The decision is made before the well is started so that the total depth can be stated on the permit. If an oil company does decide to drill deeper, it can apply for another permit from the state in order to drill to a deeper depth.

Generally speaking, because of the economics involved, the shallow formations are produced first in a new producing area. After all of the oil has been found in the shallow formations, someone will eventually drill deeper looking for new and deeper production. If a new producing horizon is found, then most all future wells will be drilled to this new depth. When all the oil has been produced from this horizon, then someone drills deeper looking for still deeper oil

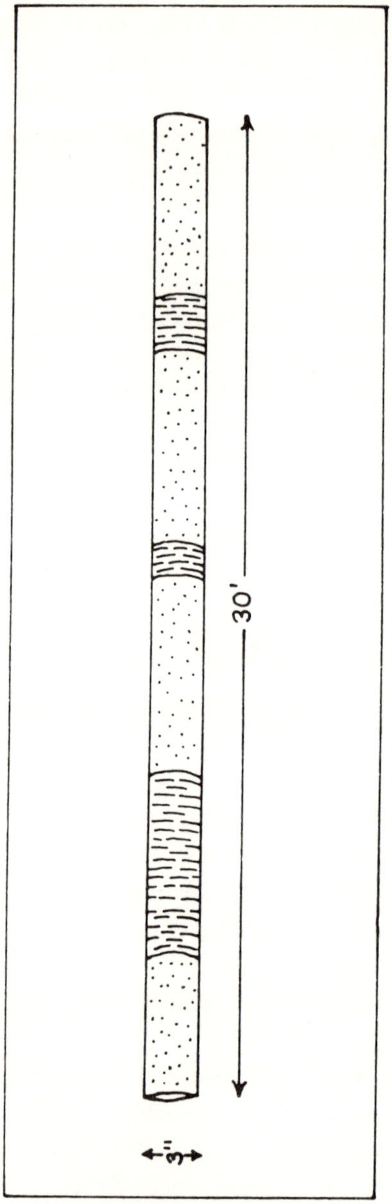

CONVENTIONAL CORE

Fig. 21—By using a special drilling bit that does not grind up the formation into small particles, a large section of the formation can be brought to the surface for examination and laboratory analysis. If an oil sand is cored, the oil can often be seen by the naked eye. There is also a strong oil odor.

producing formations. This cycle is repeated time and time again when new producing areas are discovered.

Total depth is reached . . . will the well produce?

Now that the well has been drilled to the planned depth, the drill pipe is pulled out leaving the hole with just the mud and the casing protecting the fresh water sands. Now we are ready to evaluate the well to see if it will produce oil or gas in commercial quantities. This process of evaluation is referred to as "logging" the well.

A service company that specializes in logging a well (there are several) is called to the well. The oil company that drilled the well does not log the well. This is a highly specialized phase of the oil business and the company would rather hire the services of someone who does this type of work exclusively. Logging a well consists of lowering an electrical instrument to the bottom of the well by means of a cable and then pulling it back up the hole, slowly past the formations. As the instrument is pulled upward past the various formations, electrical impulses are recorded at the surface in the form of a graph. A geologist examines the graph (log) and can distinguish the different formations of sand and shale. Sometimes this log shows whether there is oil or gas present in the sands. This is not a conclusive method of knowing if oil or gas is present, but it does give an indication. (Figure 22)

After studying the log, any sand that looks promising is further examined by "sidewall cores." This is another method of actually getting a look at a piece of a potential oil bearing formation. This sample is obtained by lowering a "sidewall core gun" into the well and stopping it opposite the formation to be "cored." A small hollow metal cylinder is fired into the formation by an explosive charge. Part of the formation is jammed into the cylinder which is still

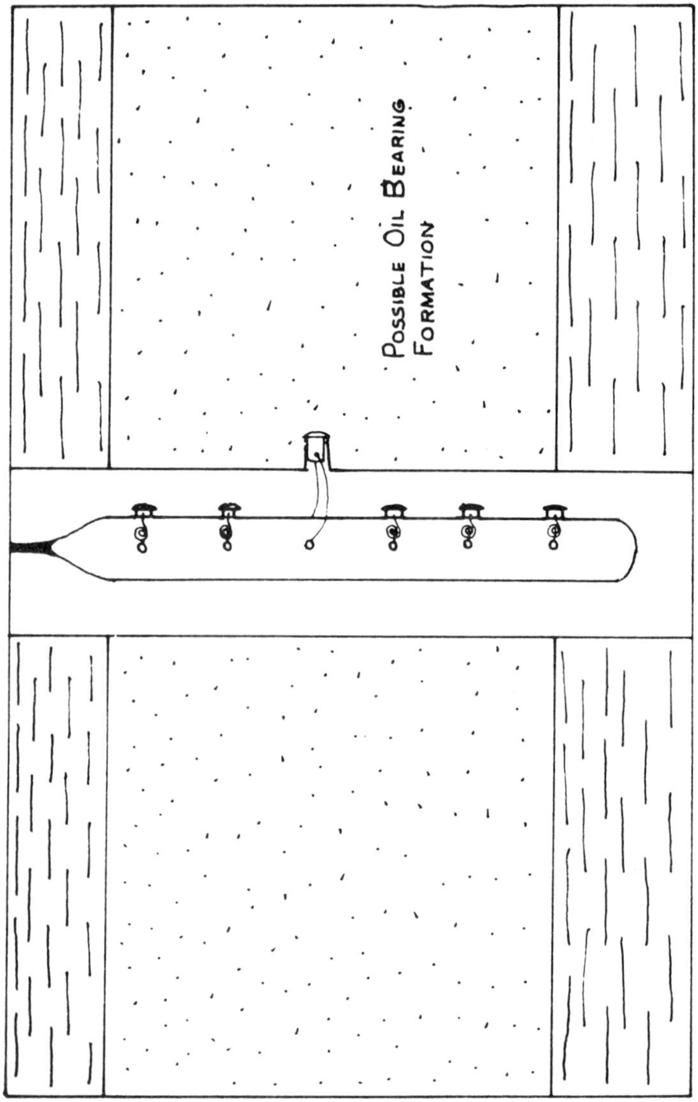

SIDEWALL CORE GUN

Fig. 22—If a particular formation looks as if it might have oil in it by examining the electric log, further testing can be done. A sidewall core gun is lowered into the hole and a sample of the formation is brought to the surface by fireing a "bullet" into the formation to be examined. This core is about 2 inches long and one-half inch in diameter. The larger conventional core is more desirable, but much more expensive to obtain. Each type of core has certain advantages that the other doesn't have.

connected to the "gun" by small cables. When the "gun" is lifted upward in the well, the cables pull the cylinder from the formation. The piece of formation that is contained within the cylinder is brought to the surface for examination by the geologist. If there is oil or gas present in the sample (core) it is analyzed at a core analysis laboratory; if the analysis is favorable, the well will probably produce. (Figure 23)

Will an oil company deliberately plug and abandon (cap) a well that they know will produce in order to come back later and produce it?

I would guess that the two greatest misconceptions about the oil business are: (1) Oil is found in a pool or a river. (2) An oil company will plug and abandon a well that they know will produce oil in order to come back years later and produce it. I hope that by now the first misconception has been dispelled and that by the time you finish this book, the second will have suffered the same fate. **This common belief is not true**, but I can understand how this idea could have come about. There have been cases where a well was declared dry and then later, was made into a producing well, but I can assure you that it was by accident and not by design. The reader should understand that there are possibly hundreds of sands present in some deep wells. By studying the log, many of the sands can be recognized as salt water sands and many can be recognized as oil, or gas bearing sands, if such are present. On the other hand, many sands fall into a doubtful class. The geologist takes sidewall core samples from each sand that looks promising, but since there are maybe hundreds of sands present, it is almost impossible to take a sidewall sample from every sand in the well. And since the log is not a conclusive instrument, sometimes an oil or gas sand just does not show up favorably on the log. A core

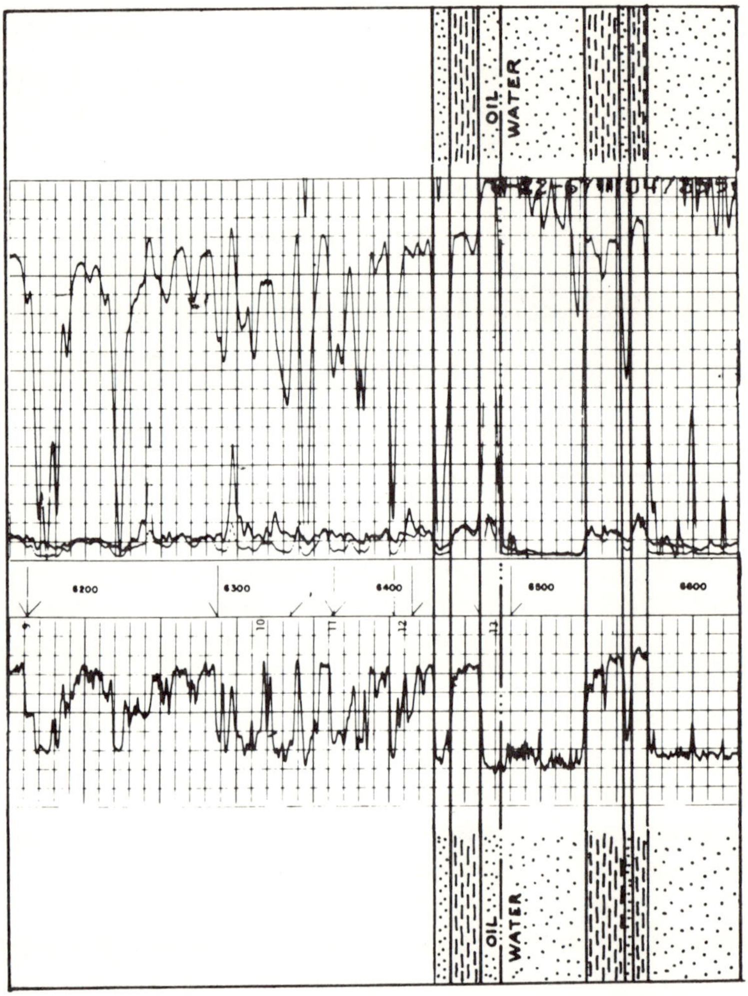

ELECTRIC LOG

Fig. 23—The top part of the illustration shows how the log actually looks. Each wiggle of the lines tells the geologist whether he is looking at sand or shale, and perhaps if there is any oil present. The bottom of the illustration shows the relationship of the electric log to the formation it records. For a clearer understanding, I have drawn in what the formation looks like opposite the curves on the log.

is not obtained from the formation and the well is a "dry hole." Some time later in another well close by perhaps, this sand looks more favorable on the log. It is tested and found to be productive. Then the log of the "dry hole" is re-examined and a decision might be made to go back into the well to test the "overlooked" sand. Sometimes it does produce . . . thus comes the idea that the well was originally "capped" on purpose.

There is another point that proves this is not done on purpose. It is not always the company that originally drills the well that comes back later to produce it. There was an oil field that had several producing wells in it for many years. Finally, the field was depleted, and all the wells were plugged, abandoned, and forgotten. Some time later, a farmer noticed that gas was seeping up in the area where one of the producing wells had once been located. One thing led to another, until finally another company became interested and decided to re-examine the logs of the abandoned wells.

There were no known gas sands present in any of the wells that had produced oil for many years; thus it was a mystery as to where the gas was coming from. It so happens that on an electric log, an oil or gas sand looks very similar to a fresh water sand. This second company went back into one of the old wells, tested what looked like a fresh water sand and found a very excellent gas sand. It was present in the same zone as the fresh water sands; thus it was assumed to be a fresh water sand; therefore, it had been overlooked originally.

This rarely happens today, however, because logging instruments are much improved and give more detailed information about every sand in the well.

As a last word on this subject, an oil company will not spend the time or the amount of money that is necessary to drill a well and then knowingly walk off and leave a well that will produce oil of gas. It just wouldn't make sense.

How is oil produced?

Now that the oil has been found, only half the task of producing it has been performed. Most people do not realize this, but getting the well into maximum production is almost as difficult as finding the oil.

A petroleum engineer takes over the task of getting the well into production. He utilizes his own skills plus those of many service companies. These companies supply the equipment and tools necessary to complete the well. The completion technique is a highly technical operation and determines the future performance of the well.

The first thing that is done to produce oil is to cement production casing all the way to the bottom of the well. This casing is similar to the casing that was used to protect the fresh water sands but it is smaller in diameter. It is cemented in the same way and generally for the same purpose. Rather than sealing off the fresh water sands, however, it seals off the salt water sands. He wants the salt water to stay where it is . . . in the salt water sands. All attempts are made to keep an oil well from producing salt water with the oil. The more salt water that a well produces, the less oil it produces.

Once the production casing is permanently cemented in the well, an even smaller temporary pipe, called a "liner" or "tubing," is put inside the casing. Eventually, holes are worn in the tubing, but it is not a serious problem. The tubing is temporary and can be easily replaced. If tubing were not used, however, and the casing had holes worn in it, the problem would be serious. Since the casing is permanent and cannot be replaced, the well could be ruined by an invasion of salt water.

Now that the casing has been cemented in the well and there is a firm bond between the casing and the formation, how does the oil get through the cement and casing and into the well? This is accomplished by using a perforating gun. It is lowered by a cable down the inside of the tubing until it is opposite the formation to be produced. By means of an electrical impulse at the surface, a type of charge is exploded in the perforating gun. By a combined explosive and burning action, holes are made through the casing, through the cement and into the formation. These holes are called perforations. (Figure 24). There are usually two to six holes per foot and they extend about a foot or two into the formation. The oil then comes into the well through the holes or perforations and either flows or is pumped to the surface.

A well can be ruined at this point

The top of the oil bearing formation is usually perforated since the bottom of the formation very often has salt water present. If a well is allowed to produce too much oil too fast, it is possible to pull the water into the well instead of the oil. Water will flow more easily through the sand than the oil will, so once the water starts flowing into the well, it is impossible to stop it. The well will never produce satisfactorily and will have to be abandoned prematurely. (Figure 25)

If the well has to be pumped, there is a pumping unit installed at the surface. If the well flows oil or gas naturally, there is just a mass of valves and pipe connections at the surface. These valves and pipes resemble the branches of a tree and is called a "Christmas Tree." As the oil comes out of the ground it is run into a special tank that separates the oil from any salt water that might be produced along with the oil.

What happens to the salt water that is produced along with the oil?

Disposing of the salt water is a very important consideration in producing oil from a well. A well may produce only oil at first, but most wells start producing some salt water along with the oil sooner or later. The salt water cannot be put into streams or allowed to run onto the ground

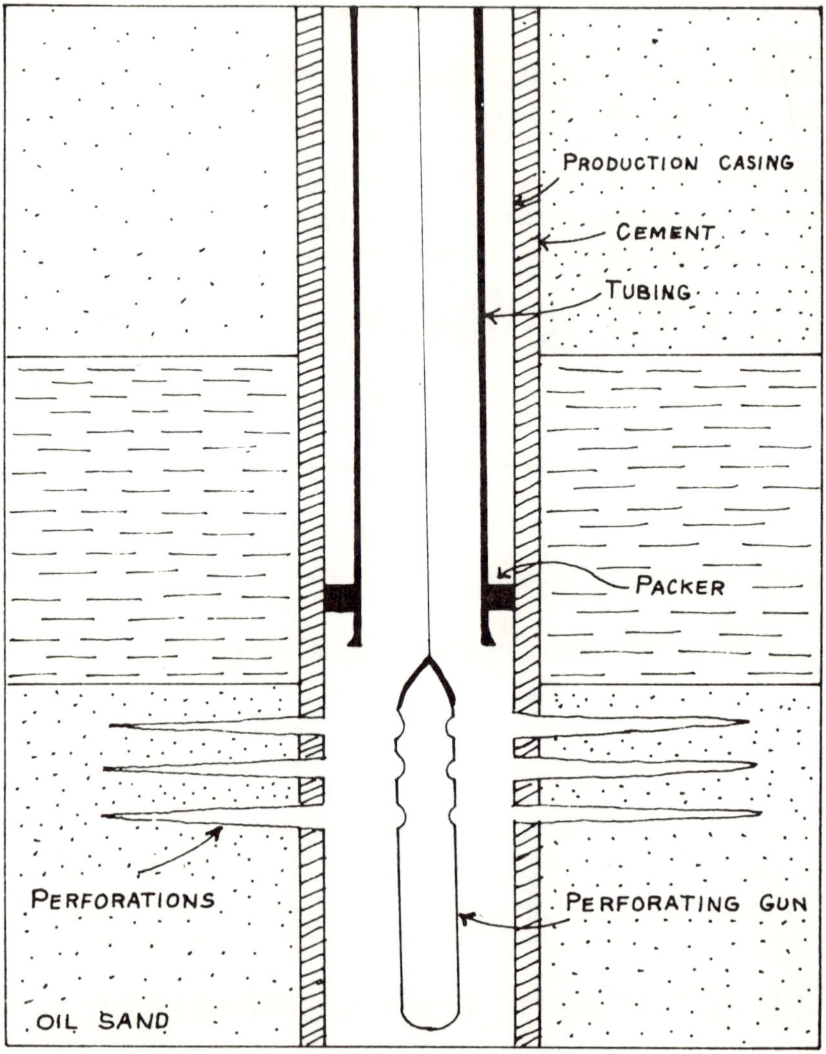

BRINGING IN THE WELL

Fig. 24—A perforating gun is used to "burn" holes through the casing, through the cement and into the oil bearing formation. If the oil will flow by natural pressure, it flows into the well through the perforations, up to the surface and into storage tanks. This is referred to as "bringing in a well."

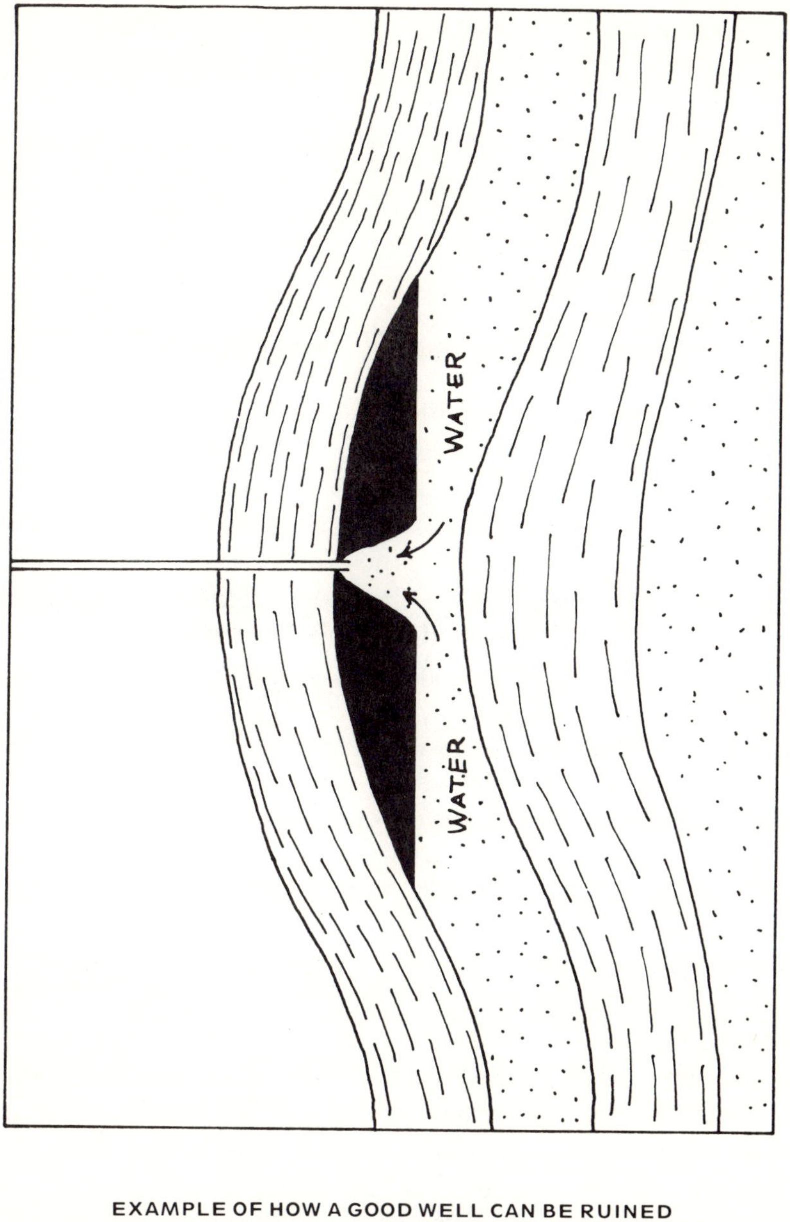

EXAMPLE OF HOW A GOOD WELL CAN BE RUINED

Fig. 25—Very often a company will ruin a good well by trying to produce too much oil too fast. Since water will flow through the sand more easily than oil, the water will "cone" up around the well perforations. Eventually the well will be producing nothing but salt water and most of the oil that should be produced is left in the ground.

because it will pollute the stream and kill the vegetation.

The salt water is usually disposed of by the use of a "salt water disposal well." A well is drilled through the fresh water sands to the first thick salt water bearing sand. The casing is perforated opposite the salt water sand and instead of producing salt water from the sand, salt water is pumped down the well and out into the salt water sand. The salt water from several oil wells can be disposed of in this one well.

How long will a well produce?

There is no exact number of years for a well to produce. . .it varies considerably from well to well. It is easier to predict how much oil a well will produce rather than how long it will produce. The amount of oil that a well will produce depends upon the thickness of the sand, the location of the well within the reservoir and several other factors. Very generally speaking, if a well is drilled on 40-acre spacing and has 10 feet of oil sand present, it will produce about 100,000 barrels of oil . . . 20 feet will produce 200,000 barrels, etc. Usually the slower a well is produced the more oil it will ultimately produce. A well might be capable of flowing several hundred barrels of oil per day, but if allowed to do so, it would probably leave a lot of oil in the ground that should have been produced. As explained before, this unrestricted flow rate damages the formation by pulling in the water prematurely or by losing too much natural formation pressure too soon.

Because of this potential damage to the reservoir, each State Oil and Gas Board will usually set a maximum amount of oil that can be produced per day from each well in a particular field. The maximum varies from field to field but it is very rare for a well to be allowed to produce over 500 barrels per day. Before anyone gets another misconception—that all wells will produce this much oil—you should know that the average production for a well in the United States is approximately 15 barrels of oil per day.

What is considered a good well?

As to how much oil a well must produce to be considered a good well, it is basically a matter of economics. If a well costs $50,000.00 to drill and produces 50 barrels per day, it is a good well. If a well costs $500,000.00 to drill and produces 50 barrels per day, it is not a good well.

Is all the oil recovered out of an oil reservoir?

Unfortunately, the answer is no! In some cases, for every 100 barrels of oil present in the reservoir, only 15 barrels are produced.

There are basically two types of oil reservoirs: one is called a water-drive reservoir and the other is called a gas-expansion reservoir. The water-drive reservoir is the type shown in Chapter 3, where the oil separates and is "floating" on top of the water. The water pushes the oil along, in a manner of speaking, and "washes" the oil from around the sand grains and into the well. This is the most efficient type of production, but even so, only 40 percent to 50 percent of the oil in the reservoir is recovered.

The gas-expansion reservoir is sealed on all sides and does not have the water drive necessary to push the oil and "flush" the reservoir. The oil and gas are mixed together, and the gas pressure pushes the oil out of the reservoir and into the well. This type of reservoir produces only about 15 percent of the oil that is present. Fortunately, there is a way of getting more oil out of this type of reservoir. A method referred to as "secondary recovery," or "water-flooding" is employed. After the wells have produced all they can by natural means, water is injected into some of the wells and an artificial water drive is devised. The water is pumped into the reservoir through these wells and as it flows through the reservoir, it "flushes" additional oil from between the sand grains and into the wells at the opposite end of the reservoir.

The oil and water are produced together and then separated at the surface.

Another 25 percent to 35 percent of the oil left in the ground can be produced by this artificial water drive. (Figure 26) Unfortunately, even after this process is employed, approximately 50 percent of the oil is left in the reservoir. There are chemicals that could be put into some of the wells that would flush even more of the oil from the reservoir, but the cost of the chemicals would be more than the income from the oil that would be produced. If someone could devise an economical method of producing the oil that is left in the reservoir, he could make a fortune.

Since some of the wells are used to inject the water and some wells are used to produce the oil in a water flood project, the question naturally arises as to who will get the oil payments for the oil produced. Before a field is water-flooded a meeting of all the landowners and all the companies that have wells in the field is held. It is agreed that everyone will receive oil payments in proportion to the number of acres he has in the field. Which wells are used for injection and which are used for producing the oil does not matter. This agreement is called "unitization" and the field is said to be "unitized."

Is all oil the same?

The grade of oil is measured by the term "gravity." A very thick black oil that is almost like asphalt is used for surfacing roads. The lowest grade oil that can be produced is approximately eight degree gravity oil. If it is any thicker than this, it is too thick to produce. This is the lowest grade of oil and also brings the lowest price. A barrel of this oil sells for about $2.00 per barrel. As the oil gets lighter the gravity gets higher. A high gravity oil such as 40 to 50 degrees will bring over $3.00 per barrel. If the oil gets any lighter, it is referred to as a condensate and looks very much like kerosene. If it gets any lighter than 60 degrees, it is no longer

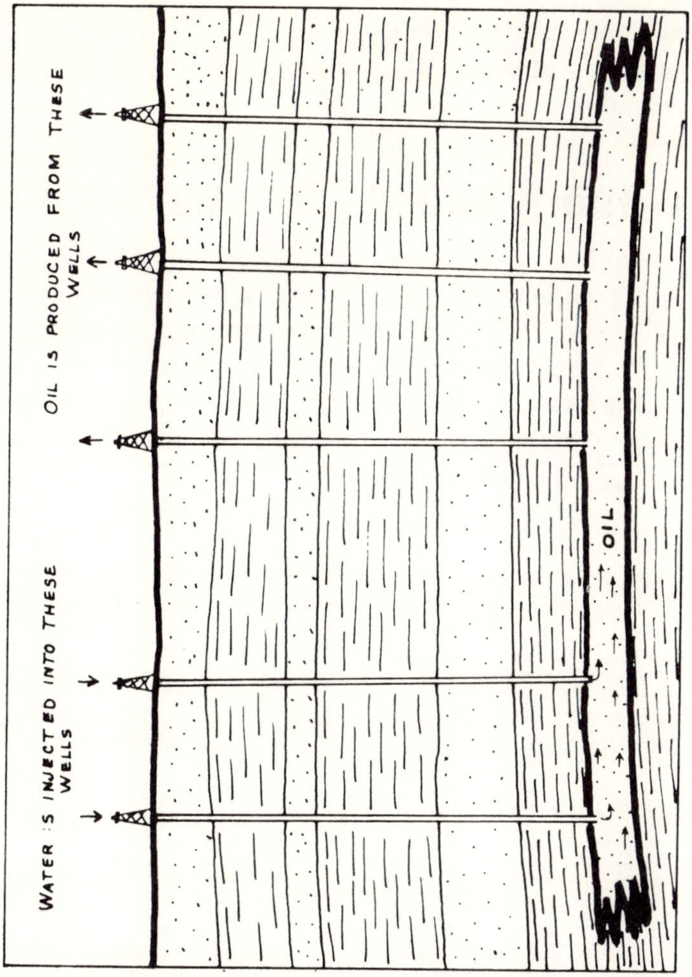

SECONDARY RECOVERY OR WATER-FLOODING A FIELD

Fig. 26—The oil reservoirs illustrated before have been water drive reservoirs, in which the water in the formation pushed the oil into the well and in so doing, flushed much of the oil from between the sand grains.

This is a gas-expansion reservoir, in which there is no water "pushing" and "washing" the oil from the reservoir. The oil that is produced is done so by the natural expansion of the gas mixed in with the oil. Only about 15 percent of the oil in the reservoir is produced when the wells stop producing. In order to get more oil from the reservoir, the field is "water flooded."

After making a study of the field, certain wells are selected to be "injection" wells and certain others are selected to be "producing" wells. Water is pumped down the injection wells and as it moves through the reservoir, it pushes oil into the producing wells. Even by this method, approximately 50 percent of the oil is left in the reservoir and never recovered.

a liquid, but a gas.

Generally, but not always, the heavier or lower gravity oil is found at shallow depths and the lighter oil is found at deeper depths.

How many gallons in a barrel of oil?

We have been talking in terms of barrels of oil instead of gallons throughout this book. You might be interested in the breakdown of a barrel of oil. There are 42 gallons of oil in a barrel. Out of a barrel of oil we get approximately 20 gallons of gasoline and 12 gallons of fuel oil used for heating purposes. The remaining 10 gallons make kerosene, greases, lubricating oils, insecticides, etc.

Chapter 7

Pass the sodium chloride please

If you were seated at a dinner table and someone asked you to pass the sodium chloride, "naturally" you would pass the table salt. "Everyone" knows that sodium chloride is just the chemical name for salt. What you might not be aware of, however, is what table salt has to do with oil and gas production. Because of its importance to the production of oil and gas, an entire chapter is devoted to salt.

Much of our salt comes from salt domes, and salt domes are closely related to some types of oil reservoirs. These domes have been mentioned in Chapter 3, but you might be interested in knowing how they're formed.

In some of the ancient seas discussed also in Chapter 3, the water was exceptionally salty. As the water slowly evaporated, an abnormal amount of salt residue remained in the sea. When the water finally evaporated completely, a solid mass of salt was left behind that was hundreds of feet thick.

An example of this process is Great Salt Lake in the state of Utah. The lake is only a fraction of the size that it was originally, and it will eventually dry up completely in the geologic future. The Bonneville Salt Flats, which border the Lake, was once a part of Great Salt Lake. Today it is a solid salt bed.

Another ancient sea later covered the salt layer and thousands of feet of sediments of sand and shale were deposited on top of the salt. Salt is a solid as opposed to a gas or liquid, but when it was subjected to the pressure of thousands of tons of sediments it became almost like putty and actually flowed. If there were any sort of structural weakness or a fracture zone in the over-lying sediments, the salt flowed up the fracture and mushroomed up through the formations, cutting through some and merely arching some that it did not reach. (Figures 27 and 28)

When this movement of salt occurred, three types of oil traps were formed: (1) traps against the side of the salt, (2) faults that were caused by the upward movement of the salt and (3) traps formed by the arching of the formations above the salt.

Besides being important for oil production, salt domes are important for other reasons:

(1) Table salt

Some salt domes push up very close to the surface of the ground. When they do, salt can be mined in the same way that coal is mined. If the salt is pure enough for human consumption, it is processed and made into common table salt. If you have never taken a close look at the salt you put on your food, you might be surprised to learn that each salt grain is a perfect cube. No, the salt company did not shape the grains this way . . . this is the natural crystalline structure of salt. It is mined in huge chunks and as it is crushed in processing, it breaks into cubes naturally. If a salt dome has too many impurities in it to be used for human consumption, it is mined and used primarily for melting ice on streets and walks.

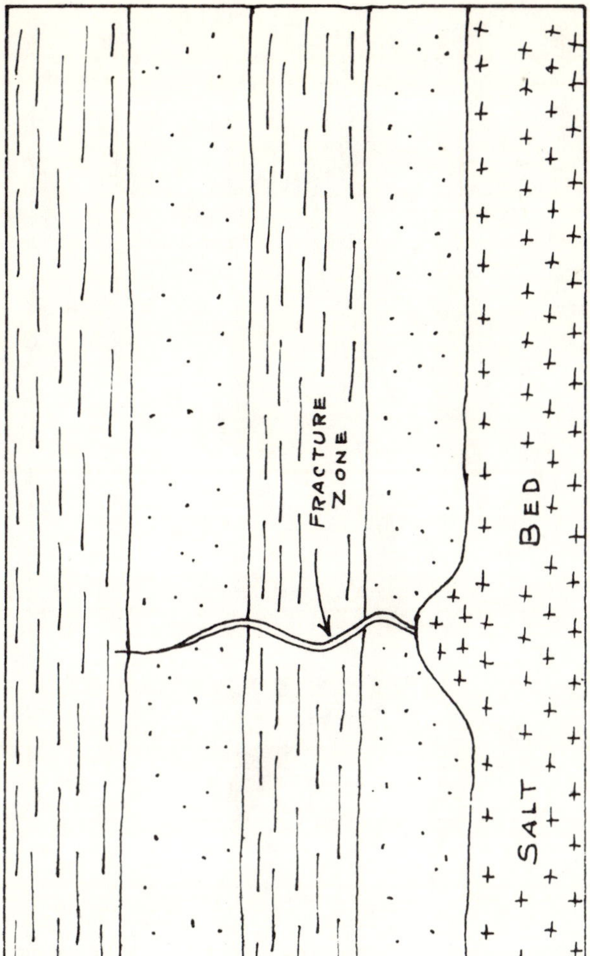

BEGINNINGS OF A SALT DOME

Fig. 27—This illustration shows the original salt bed with a portion of the salt starting to flow into a weak fracture zone in the overlying sediments.

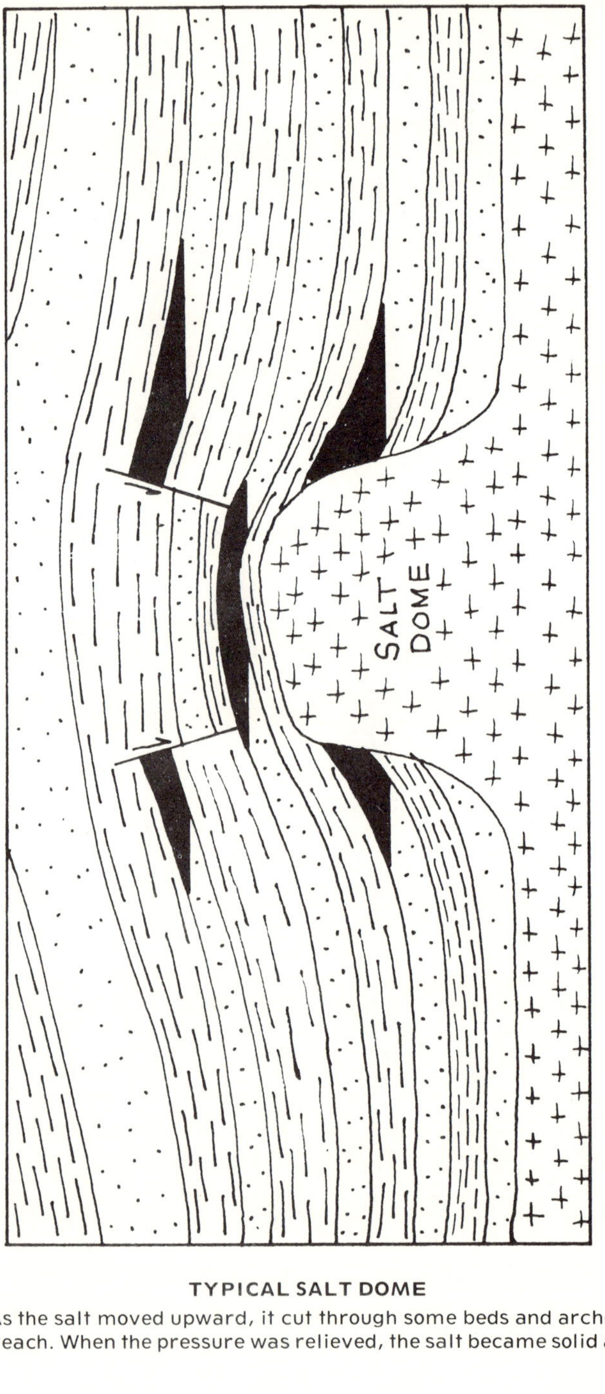

TYPICAL SALT DOME

Fig. 28—As the salt moved upward, it cut through some beds and arched the ones it did not reach. When the pressure was relieved, the salt became solid again.

(2) Sulphur

Some salt domes have a layer of rock on top that is called "caprock." The caprock is composed basically of limestone, but often has sulphur present within the limestone. Instead of mining the sulphur by the same method used in mining salt or coal, a more practical method has been devised. Wells are drilled into the caprock and then hot water is pumped down the well and out into the limestone containing the sulphur. The sulphur is melted by the presence of the hot water and thus becomes a liquid. It is then pumped to the surface where it is allowed to cool and turn again into a solid. (Figure 29)

(3) Gas storage

A huge cavity is washed out of the center of the dome by forcing water down a well and dissolving a large part of the center of the dome. Natural gas is then placed in the dome and it serves as a huge storage tank until the gas is needed.

(4) Atomic energy tests

The Atomic Energy Commission has conducted certain types of experiments within a salt dome near Hattiesburg, Mississippi. A cavity was hollowed out of the center of the dome and a small atomic explosion was set off. The tests were conducted for many reasons, two of which were (1) to see if heat energy generated by the explosion could be stored within the dome for a length of time and then used at a later date and (2) to try to determine if secret underground atomic tests by foreign countries could be detected.

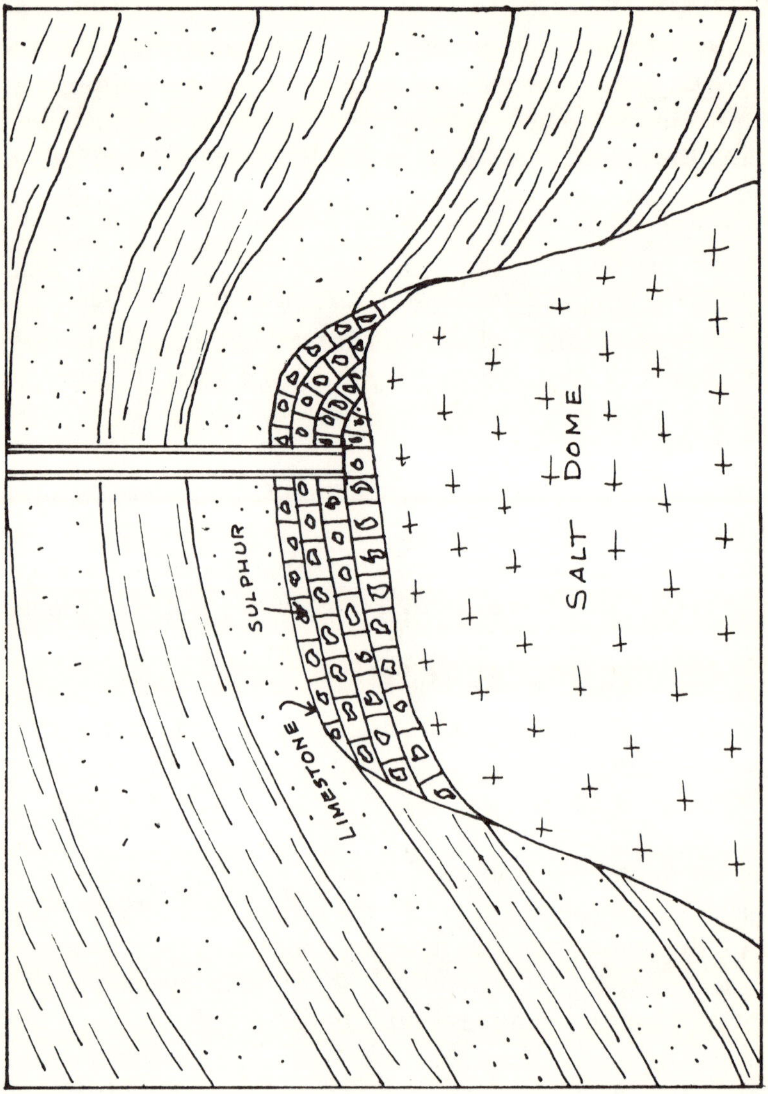

CAP ROCK AND SULPHUR

Fig. 29—Very often, a salt dome will have a cap rock located at the crest. The cap rock is basically composed of limestone but sometimes sulphur is associated with it. If the sulphur is plentiful enough, it is mined by a hot water process. Hot water is forced down a well, melts the sulphur, and forces it to the surface to be refined.

Chapter 8

Offshore drilling

Offshore oil wells have made the news quite frequently during recent years. There have been oil spills off the coast of California and oil fires off the coast of Louisiana in the Gulf of Mexico. These accidents occur onshore also but they are more difficult to bring under control at offshore wells. This fact brings us to an important point: the only difference between offshore and onshore oil operations is found above the surface. There is no basic difference in the reservoirs, traps, or gravity of the oil.

Practically all of the oil fields in the Gulf of Mexico are caused by salt domes; therefore, in exploring for oil in the Gulf of Mexico the search is for salt domes. The seismic method is used primarily.

There are no individual mineral owners offshore. All mineral rights are owned by the United States Government or by the State, depending upon how far they are located from the coastline. Periodically, certain tracts of minerals are put up for bid by the Federal or State government. Oil companies submit sealed bids and the highest bid gets the lease. Large sums of money are spent during these sales and very often there is a wide difference of opinion as to how valuable a certain lease is. A particular oil company once bid seven million dollars for a lease and was the high bidder . . . the next highest bid was one million dollars. As the saying goes, six million dollars were left on the table.

All costs are approximately tripled for offshore oil operations for reasons that will be readily apparent later in this chapter.

Drilling platforms

One of the major costs connected with offshore drilling is constructing a base from which to drill the wells. Some wells are drilled from a temporary platform that can be moved from place to place. In most cases, however, a drilling platform is constructed by driving steel pilings deep into the floor of the Gulf. It is built approximately 50 feet above the water level in order to withstand hurricanes. Because of the high cost of constructing a drilling platform—as much as two million dollars—we find one of the main differences between offshore and onshore drilling. As many as 20 wells can be drilled from the same platform. The wells have to be drilled directionally by using a whipstock and directing the hole to the desired location. This is an expensive way to drill a well and is beset with problems, but it is far less expensive than building a separate platform for each well. (Figure 30)

Because of these many wells all being produced from the same platform, if one of the wells happens to catch fire, there is the danger of the others catching fire also. This is the situation that has occurred in the Gulf of Mexico several times. It is a rare occasion when a well catches fire but the publicity that it receives leads the public to think that it is a common occurrence. When and if it does happen, the fire or "blowout" is extinguished in the same way that it is done on land but it is a much more costly operation. Mobility on water is much more difficult than on land; thus the procedure takes

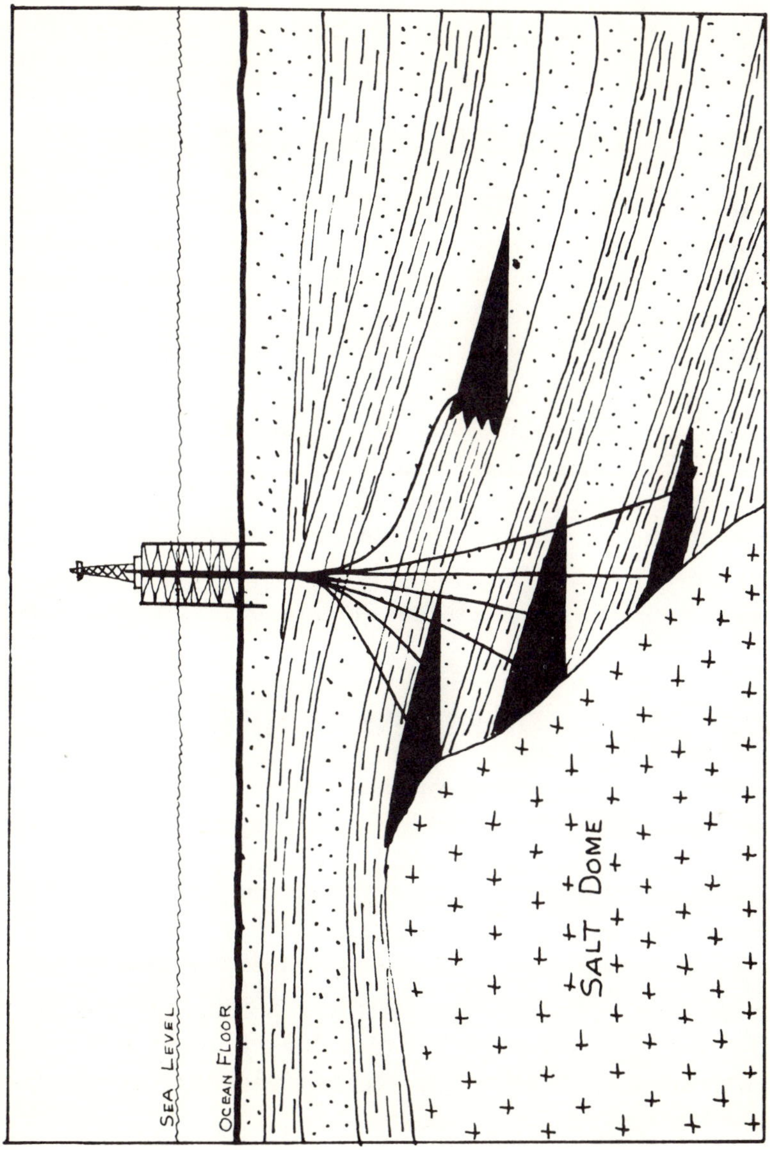

MULTIPLE WELL OFFSHORE DRILLING PLATFORM

Fig. 30—All wells on an offshore platform are drilled vertically to a depth of about 3000 feet. At this point, a "whipstock" (mentioned earlier) is used to drill the wells at an angle to the desired location. There can be as many as 20 or more producing wells on one platform.

longer. The inherent danger in this type of operation also requires slower and more deliberate work.

Living conditions for workers

Another major difference between offshore and onshore drilling is the working and living conditions of the men drilling the well. Some of the wells are as much as 70 miles offshore so the drillers and roughnecks usually stay at the drilling platform for 10 days at a time and then are off duty for four days. Some platforms have a large vessel anchored next to the drilling platform and is connected to it by a gang-plank. Some platforms have the living quarters built on the platform. In either case there is a cook aboard, the food is good, there is a recreation room with television, and the fishing is excellent. All in all, the living conditions are as good as can be expected and in some cases, better than can be expected.

Future of offshore drilling

With the oil consumption in this country growing tremendously every year there is an urgent need to discover new oil reserves. It is highly possible that these needed reserves will be found by offshore drilling. We hear a lot of criticism about offshore drilling today and many rules and regulations are being put into effect, implying that oil companies are insensitive to the problems of conservation and pollution. I feel confident that the oil companies will do whatever is necessary to protect our natural resources both under and above the surface.

Chapter 9

Risks and rewards

Even though the preceding chapters have dealt with most aspects of the oil business in a very general way, by now you should have a working knowledge of how oil is formed, found and produced.

This chapter will deal in detail with another phase of the oil business: selling and buying leases, minerals, and royalty . . . and investing in a drilling venture. Just as in the preceding chapters, there are also misconceptions about this phase of the oil business. Some people think that anyone who has an interest in a producing oil well has his fortune made . . . others think that anyone who invests in the oil business is just throwing money away. Somewhere in between these two extremes is the truth of the matter.

If a person fully realizes the risks involved in finding oil and fully realizes the rewards he will enjoy if a well produces, he can then make a knowledgeable decision about whether to invest. Both sides will be explained in the following pages.

Would you accept a producing oil well as a gift?

Before you answer . . . don't answer . . . until you are made aware that it could cost you dearly to own a well. This subject was mentioned in Chapter 2, but it is a point that cannot be overemphasized. After a well has produced for several years and the oil production has dropped off . . . and the well is making oceans of salt water . . . and the equipment is shot . . . and the tubing in the well needs replacing . . . the amount of oil that the well is producing will not even pay for the high cost of producing it. Of course this is not the usual situation, but it does make my point . . . it's not only how **much** oil production an individual or a company is receiving, but also how much **cost** is involved in getting this production.

This chapter should be of special interest to you landowners who have not leased your land. In the event that you have an opportunity to lease, you will be familiar with the terms, conditions and rights of both parties, the buyer and the seller. It is possible that even landowners who have already leased will also find the information enlightening, for very often a person will sign a lease without knowing fully what is contained within.

As the title of this chapter, **"Risks and Rewards,"** implies it should also be of interest and possibly financial reward to anyone wishing to invest in the oil business.

In order to participate even in the slightest way, it is absolutely imperative for an individual to be familiar with the basic types of legal instruments, oil and gas leases, royalty deeds and mineral deeds.

Oil, gas and mineral lease

A landowner owns all the minerals, including oil, gas, coal, sulphur, gravel, uranium, gold, etc., that are present beneath the surface of his land unless of course he has sold his mineral rights. In referring to "landowner" from now on, it is assumed that he still owns all the minerals under his land. I would also like to point out that it is perfectly legitimate for an individual to buy a lease for himself. He may be planning to drill a well himself or he might be planning on buying up a block of leases for an oil company. My reason for making this point is that some landowners do not want to sign a lease unless the lease is made out to an oil company.

When an oil company leases your minerals (land), it is buying the right to drill a well on your land in search of oil. The company will pay you for the lease and you will also receive a share of any oil that might be found. This share of the oil that you receive is known as **royalty**. When a company buys a lease from you, it will pay a certain lump sum price referred to as a **bonus**. This bonus will buy the lease for one year, with an option to renew. If oil or gas production has not been found on your land within one year, the company can extend the lease for another year by paying you a rental fee which is usually one dollar per acre. This is a point that is confusing to some landowners; they are under the impression that they will receive the same amount of money every year in rental fees as they were paid when the lease was first bought. The first time they become aware of this is at the end of a year; they are expecting a check for a much larger sum than they actually receive. Some feel that they have been misled when this happens, but they should not feel as though it was done on purpose. The terms are very clearly spelled out in the lease and, I might add, not in small print.

If the term of the lease is for a length of ten years, which is more or less standard, the oil company can continue to pay a rental fee of one dollar per acre every year and keep the lease in force for the ten year period. The oil company can cancel the lease at any time by not paying the rental fee on time but the landowner cannot cancel for the entire ten year period. This may seem to be an unfair arrangement for the landowner, but you must remember that the company is putting up the money. If a company buys a ten year lease from you, naturally you are anxious for it to drill on your land as soon as possible. If the company does drill during the first year and drills a dry hole, it would not be fair to have to continue paying rental fees for another nine years . . . this company would probably cancel the lease by neglecting to pay the next rental fee that was due.

On the other hand, I met a landowner not very long ago who was trying to cancel a lease. An oil company had bought a lease from him a couple of years ago for five dollars per acre, but oil had been discovered near by since then and the prices had jumped to twenty-five dollars per acre. He figured that if he could cancel his lease, he could then re-lease his land for twenty-five dollars per acre. He tried to cancel by not accepting the rental fee when it was due. This does not work though, because when you sign the lease you usually name a bank to which the rental payments are sent each year. The deposit is made automatically and you have no choice as to whether or not you will accept the rental payment.

In the event that oil or gas is discovered on your land, the rental payments every year are not necessary in order to keep the lease in force. The company will own the lease for as long as oil or gas is being produced, even if it is longer than the original ten year term.

The amount of royalty that the landowner usually keeps is 1/8 of all the oil or gas that is produced. It is interesting to note that the lease provides for the landowner to actually keep one barrel of oil out of every eight produced. Fortunately, the oil company will sell all the oil and send the landowner a check for his share every month. The oil company keeps 7/8 of the oil but remember that they may have spent hundreds of thousands of dollars to find and produce the oil. The landowner gets his 1/8 free of all costs.

Under certain conditions, the landowner might retain more than 1/8 of the oil. Any amount of oil retained that is more than the normal 1/8 royalty is referred to as overriding royalty.

Language and terms of the lease

There are many sections to an oil lease but the following are some which confuse the landowner most frequently:

1. Lessee is the one buying and lessor is the one selling.

2. The amount of money paid for the lease need not appear on the lease to be legal. The lease is recorded for public information in the County Courthouse and you may not want the amount to be shown on public record. A sum of ten dollars is usually inserted to make the transaction legal.

3. A pooling clause merely means that an oil company can form a drilling unit and include more than just one landowners land in the unit. They can "pool" your land with someone else's land. This clause works to the advantage of the lessor and the lessee because a well can be drilled at the most favorable location from the geological viewpoint.

4. A well cannot be drilled within 200 feet of a building or a high power line. (This rule varies by states.)

5. When an oil company drills a well on your land they can use existing roads on your land or water from your rivers and ponds. They cannot use water from your fresh water well however, unless you give permission.

6. As a matter of good will an oil company will usually pay for any crops that have to be damaged during the course of drilling a well.

How leasing is done

Think back now to Chapter 3 dealing with the oil companies locating the ancient seas, or basins. These areas are where they expect to find oil fields. Once an oil company locates a basin, it will then buy leases over a very broad area with the anticipation of exploring further to try to find specific places to drill. The company knows that most of the leases that it buys in this manner will not be productive, but it buys over a broad area because knowledge at this point is limited as to where the individual oil fields might be located. When leases are bought at random like this, the oil company will set a maximum amount to pay for leases and will buy all it can at this price. If someone wants more money for his lease than the maximum amount set by the company the lease is not bought. The person buying for the company could not pay more than the authorized price.

After the area is studied by the oil company by one of the methods mentioned earlier, several places that look as though there could be prospects for an oil field might be chosen. Then the company becomes more selective in buying leases. Chances are that the company already has bought some leases in this area during the random buying of leases. If the company thinks highly of the prospect, a higher price for any leases not bought the first time around will be paid now. The company does this in order to get enough acreage in one place to drill a well. An oil company could not be expected to drill a wildcat well in an area in which it did not own all or most of the leases. The reason being, if it is willing to spend the large amount of money necessary to discover a new oil field, it certainly wants to have the surrounding leases in order to be able to drill the subsequent wells.

You might now be thinking that the best thing to do is not lease your land at first—just wait and the price will go up! This is not necessarily what happens, however. This type of reasoning could cause you to lose the opportunity to lease at any price. After a study of the basin is made, the oil company wants leases located only where there might be oil. So if you did not lease the first time around, it might not want to buy your lease at any price after the area has been studied. So it boils down to this: if you don't lease the first time around and an oil company does later want to drill close to your land, you will probably get a higher price; but if you don't lease and a decision is made not to drill close to your land, the company probably will not buy your lease at any price.

I once represented a client who was offered $15.00 per acre for a lease on his land. He asked my advice as to whether to lease for this price. I checked the area and found that $15.00 was the going price, and a fair price I might add, since there had never been any oil production close by. I explained the situation to him as I have just explained it to you in the preceding paragraph, and advised him to lease. He said he would rather take a chance that the company would find a prospect on his land and pay him more for his lease later. The company never came back to see him so he asked me to contact the company that made the offer that he would be willing to accept the $15.00 per acre. I contacted the company and received a very nice letter saying that it had reviewed the acreage and was not interested in buying any more leases in this particular area. I then contacted some other companies for him to see if they were interested but I could not sell the lease at all.

Another point to keep in mind . . . if an oil company doesn't get a lease, it surely cannot drill a well on your land. Some landowners have refused to lease for any amount of money— they say they don't need the money and don't want to. They give no reason for this attitude, but I suspect that often they do not understand what a lease is and for that reason would rather not sign anything. If an oil company runs into enough people in one area who will not lease, they will not buy any leases at all. They will simply move to another prospective area where they can buy leases. When a company encounters this type of problem, both the oil company and the landowners lose.

I heard a landowner once say that an oil company "cheated" him because he was paid $15.00 per acre and then two years later, the same company came back and paid a neighbor $25.00 per acre. This is an unfair statement because during this two year period, oil was discovered nearby. This always causes the price of leases to go higher. Likewise, a dry hole could have been drilled in the area and the price of leases could have dropped below $15.00 per acre. Prices go both ways!

The important thing to remember is . . . the value of oil properties does not stay constant. A house, automobile, or other tangible property will not fluctuate much from day to day, but oil property can potentially be worth thousands of dollars one morning and worthless by that night.

Mineral deed

Landowners, beware! . . . of this one. If you sell your mineral rights, you have nothing to lease and you get no share of the oil production.

Now there is certainly nothing illegal about someone selling his minerals . . . people do it every day. Unfortunately, some people have sold their mineral rights without really understanding what they were doing. In every type of business there are people of questionable character; unfortunately, the oil business has its share. I'm sorry to say that there have been cases of a person misrepresenting the facts and **buying** minerals instead of **leasing** them. To a landowner who has never seen a lease or a mineral deed this could easily happen; so be sure of what you are signing. There is probably nothing more frustrating than for a landowner to see producing oil wells all over his land and not own any of the mineral rights.

I would like to make sure that you understand that if someone wants to buy your mineral rights and tells you he wants to buy them, there is nothing dishonest about it. Just agree on a price and sign the papers.

Royalty deed

As explained earlier, when a landowner leases his land the share of the oil rights that he keeps is known as royalty. Very often a landowner will be approached to sell part of his royalty. This means that he would be selling part of any oil that might be found on his land. If he sold one half of his royalty he would be selling one half of any oil payments that he might receive in the future if oil is found. If he sells royalty, he would still receive the full rental payments every year from the company that leased his land. Selling royalty in no way affects the lease you have made with an oil company.

You might wonder why anyone would want to sell part of his royalty. As will be explained, it is sometimes a very wise decision to sell part of one's royalty.

Since there is such a wide price range in buying and selling royalty, it would not be practical to suggest what any general royalty should sell for. The following information should help you to know how much you will be paid every month if you own producing royalty. Then you might have a better idea of what it might be worth to sell it. Example: If a well produces 100 barrels of oil per day (3,000 barrels per month) your share, or 1/8, will be 375 barrels per month. If you have some use for 375 barrels of oil you are entitled to keep it. On the other hand, if you have a greater need for the money it will bring in, the oil company will sell it for you. If the oil sells for $3.00 per barrel you would get a check for $1,125.00 per month, less taxes.

As for selling part of your royalty, if your land is located a mile or so away from a producing well, it might not be worth anything or it might eventually be worth as much as the royalty under the producing well. It depends on the size of the field and on how anxious a buyer is to buy and how anxious a seller is to sell.

If you, as a royalty owner, own more than just a few acres, it is usually a good idea to sell some of your royalty. If your land turns out to be non-productive you have made some money by selling . . . if your land produces you still have enough royalty to get a nice check every month.

Oil fever

It is very easy to catch a common malady known as "oil fever" and just "know" your land is going to be productive. But all the wishful thinking in the world will not change the geologic conditions that have existed for millions of years. One of the purposes of this book is to try to make you aware of this fact. The following are a couple of examples of unwise judgement that have been multiplied thousands of times. You usually don't hear about these stories because it hurts the person involved too much to talk about it.

A well was in the process of being drilled and because of the oil sands that appeared to be productive, royalty was being bought and sold all around the well. A particular landowner owned 20 acres in the next tract to the drilling unit on which the well was drilling. In fact, his land came up to within 660 feet of the well. Since it appeared that the well would produce several hundred barrels of oil per day, someone had offered him $40,000.00 for all of his 20 acres of royalty. He asked my advice and I recommended that he sell one half of his royalty for $20,000.00 and do it quickly before the buyer changes his mind, for this was an excellent price for unproven acreage. I explained that if a well on his land were dry he would certainly have profited by selling. And if his land proved to be productive his remaining 10 acres would have paid him well. He had caught "oil fever" though, so he said, "if half of my royalty is worth $20,000.00 today, it'll be worth more later on . . . I think I'll just hold out and sell for $30,000.00." Now if this man had been wealthy I might have understood his attitude, but he could have really used the money; in fact, it was probably more money than he had ever had or even hoped to have. Well, no one paid him $30,000.00 for half of his royalty so he didn't sell. After the well was drilled, it was discovered to be separated from his land by a fault. His land did not have a drop of oil under it and he could not have then given away his royalty, because it was worth absolutely nothing. Unfortunately, he probably will never have a chance to profit from his mistake. He, like most landowners, got only one chance like that.

In another example, a wildcat well was being drilled. The closest field was 12 miles away so it was not considered to be in the middle of oil production. I studied the area in detail and felt that this wildcat had a good chance of finding oil. So I decided to try to buy some royalty close to the well in the event that oil production was found. I contacted an individual who owned about 200 acres all around the well and offered to buy some of his royalty for $100.00 per acre. This was a fair price for the area but he said he would hold out until the price got "interesting." He never would say what price he considered "interesting" but I could tell that it wasn't anywhere near the price I offered. I explained that his 200 acres, if productive, would pay him more than he could ever use . . . and if it were dry he would not come out empty-handed. He was still waiting for the price to get "interesting" when the well ended up a dry hole, as most wildcats do. I don't have to tell you what his royalty is worth now. One landowner in the area owned 40 acres and sold half of it to me. He's the only one who came out ahead.

I have heard landowners make the following statement: "If my royalty is worth that much to you, then it is worth that much to me . . . I'm not going to sell." This is not true for several reasons, but the main fallacy in this line of thinking is that the landowner forgets that the buyer is usually spreading his risks. He is buying royalty around several wells and if some royalty turns out to be worthless, he has a chance to make it up in another area. On the other hand, the landowner usually has everything riding on one well. If it's dry, he has no chance to make it up somewhere else.

Most of you are probably thinking that you would have sold part of your royalty in the two examples that I've conveyed to you. But when you see an oil derrick go up and a well is started close to your land, something happens to your sense of perspective and all kinds of thoughts of riches start popping into your mind. This is what I mean by "oil fever." It can happen to anyone who is not fully aware that any well anywhere can be dry.

Even those who are fully aware of this fact often catch this not so rare malady of "oil fever."

Types of investments

In order to invest in the oil business intelligently, it is necessary to know the few basic facts that I have tried to explain in the preceding chapters. Now that you are aware of these facts, I will try to explain how different types of investments are made.

Buying mineral rights

By buying minerals you are probably looking at a long term investment, but one that can be very profitable. I know of people who bought minerals some years ago; they have since leased the minerals several times for more than they originally paid for them. There are far less minerals sold today, however, than years ago when landowners never dreamed that oil would ever be found on their land. The price of minerals has risen to a point that you seldom find "bargains" anymore, but you can still make a good buy now and then.

Buying an oil and gas lease

You buy a lease for one of two reasons: you plan to drill a well yourself or you plan to sell it for a higher price. If you buy a lease for resale at a higher price, you must have a reason for thinking that someone else will be willing to pay a profit for it. Some of these reasons might be:

1. You might gain access to some information indicating that an oil field is located at a certain location. You would buy leases early with the anticipation that someone will want to drill in the area. If your leases are located in the right place, a company planning to drill a well will pay you a profit, plus a small share of any oil that might be found.

2. If a particular company has been unusually successful in finding new fields, you might buy leases in the same general area in which the company is buying. The leases are usually bought on the fringes of the area where the company is buying. If a field is discovered, it might turn out to be

larger than expected and the fringe leases will be potentially productive.

3. If a new producing area is discovered, you might buy leases over a broad general area with the hope that some of them will prove to be productive. This relies completely on luck and although some say they would rather be lucky than smart, this is not the wisest way to buy leases.

Rather than buying leases for resale at a higher price, some individuals buy leases with the anticipation of paying their share of the cost of drilling a well. For example, you might have bought a lease on five acres that was located on the fringe of an area where an oil company was buying leases. The company might drill a wildcat that turns out to be productive and a new oil field is discovered. As the field is developed and wells are drilled, the field may prove to be large enough to include your five-acre lease. Eventually the oil company may want to drill a well at a location that would include your five acres in a 40-acre drilling unit. You would have to pay for five-fortieths of the total cost. If the well cost $40,000.00 to drill, your share would be $5,000.00. Since this type of participation can become quite expensive, especially in a deep well, you would probably have to be in a high income bracket to be able to afford it. The cost of drilling and producing a well can run into several hundred thousand dollars.

Participating in this way also carries a certain amount of liability with it. If the well encounters problems that run the cost up, or if the well "blows out," the cost could run into millions of dollars. This type of participation offers greater rewards but the risk of having to spend more money than anticipated is also greater.

Buying royalty

Since I discussed royalty buying thoroughly a few pages back, you have probably already figured out for yourself how it works. An important point to remember is, the price that is paid for royalty around one well doesn't necessarily stay the same for all wells in the area. It depends upon the performance of the well, how thick the oil sand is and how many acres are in the drilling unit. The number of acres in the drilling unit is very important in determining the value of royalty. If you buy one acre of royalty in a 40-acre drilling unit, you will receive one-fortieth of the royalty payments. If you buy one acre in an 80-acre drilling unit however, you will receive only one-eightieth of the royalty payments. This means you would receive only half as much income each month. You can readily see that it is to the advantage of the royalty owner to have smaller drilling units. By the time the well stops producing, you will have received the same total amount of income however, because a well on an 80-acre unit is supposed to produce twice as long as a well on a 40-acre unit.

Working interest

This is the method by which most people invest in the oil business. Up to this point in the book, most all references to drilling and oil production have been made in relation to major oil companies. There are also small oil companies, known as independents, engaged in the oil business. The two are referred to as "majors" and "independents."

When a major oil company decides to drill a well, they of course have the funds available to pay for the drilling. More important, they have enough funds available to drill a number of dry holes without going broke. And they will drill dry holes; of this you can be sure.

An independent oil company may or may not have enough funds available to carry them through the dry holes that they will also drill. If they are large enough and have enough funds available, they operate just as the majors do and pay for the total cost of drilling a well with company funds. On the other hand, if a company doesn't feel as though it can take the full financial risk of drilling, it will sell working interests in the well. It will usually sell interests to other small companies—interests in the amount of one-fourth of the total cost, for example. It is not unusual for four small companies to drill a well together. Then if they drill a dry hole, it doesn't seriously affect the financial situation of any of them.

Very often an individual will drill a well in the same way that the small companies do . . . by sharing the costs with others. When an individual drills a well, he will sell interests that are much smaller than the one-fourth interests sold by the companies. He will often sell a one-sixty-fourth interest in a well, a share small enough for just about anyone to participate. This is the type of working interest that the average investor is interested in. He can buy a small interest in several wells rather than putting a large amount of money in any one well.

Because of the high cost of drilling a deep well (deeper than 10,000 feet), an individual usually drills in an area where oil production is found at a shallow depth. He will also drill in an area where many wells have already been drilled so that a geologist will have an abundance of information available to help him decide where to drill.

Once an oil prospect is located, the leases are bought. Then a drilling contractor, who is someone who owns a drilling derrick and equipment, will submit a bid for drilling the well. The total cost of mapping, surveying, leasing, legal work, drilling and logging is then totaled. If, for example, the total cost is $32,000.00, then shares are sold in order to raise money. If a person wanted to invest $1,000.00, he could buy a one-thirty-second working interest in the well. If the well turns out to be dry, this is his total cost. If, after the well is drilled and evaluated, it appears that the well will produce, each working interest owner will have to pay his share of the producing costs. This might cost another $1,000.00 per one-thirty-second interest. Now a person has $2,000.00 invested and the well is ready to produce.

If the well produces 3000 barrels per month the ⅛ royalty will first have to be subtracted. Royalty owners get their share of the oil first and working interest owners share in the remaining oil. The royalty is 375 barrels, leaving 2625 barrels per month for the working interest owners to share. Then one-thirty-second of 2625 barrels is 84 barrels. If the oil sells for $3.00 per barrel, the owner of the one-thirty-second interest will receive approximately $252.00 per month. If the well produces a total of 200,000 barrels, a one-thirty-second interest will pay the working interest owner a total of approximately $16,000.00.

As a working interest owner, he will also have to pay monthly operating costs, such as electricity to pump the well, the expense of having someone check the well every day to see if it is functioning properly, etc. It varies considerably from well to well, but the average cost is probably three to four hundred dollars per month. Of course any work that is done on the well or any repairs that are needed will be paid for by the working interest owners.

Any of the figures that I have used in the example can vary but this will give you a rough idea of what to expect from working interest.

Taxes

Because of the high risk of searching for oil plus the fact that as the oil is produced it cannot be replaced, there are certain tax allowables in investing in oil. I hesitate to use the term "tax advantage" because "advantage" implies that you have an unfair opportunity over someone else. This of course is not the case . . . the tax structure is so designed to encourage oil exploration, a very risky business.

The tax situation is so complex that I could not begin to explain it in detail so I will just mention a few of the highlights: The drilling cost ($1000.00 in the above example) is fully deductible in the year that the well is drilled. Approximately $300.00 of the producing cost is deductible also in the year that it is spent. The cost of the remaining producing equipment is depreciated over several years.

If you receive $3,000.00 per year in oil payments, you can deduct 22 percent depletion allowance, which is $660.00. You do not have to pay taxes on this amount.

If you're going to invest in the oil business, you must look upon it as an investment and not as a one-shot deal. Some people will try one well and if it's a dry hole, he's sour on the oil business and to hear him talk, everyone in the oil business is crooked. On the other hand, if the well produces, he knew it would all along because it was drilled on the south side of a creek or because Grampa said that oil would be found here some day or because he just felt it in his bones. He tells his friends that the next time he has a sure thing he'll let them in on it, etc. . .He expects everything that is drilled to be productive after he gets lucky on his first well.

It is very important for the investor to realize that most wells are dry but the rewards are great enough so that if a well hits only every once in a while, a profit can still be made.

Conclusion

I have tried not to go into too much detail in this book, nor have I tried to give exceptions to some of the statements that I have made. There are exceptions and conditions to just about any statement on any subject but to have listed any on this subject probably would have bogged you down to the point that you would understand nothing.

At one time or another I guess we have all had the day-dream of seeing a producing oil well on our land. We have always equated oil with riches but we've never known much more about oil than we've seen in the movies—an oil well wildly spewing oil all over the place. Maybe you won't ever have an oil well in your back yard but you now know the basic facts about oil production. If you ever do lease your land or have a chance to invest in an oil well, you should now be able to carry on a very knowledgeable conversation.